A TEXT-BOOK ON
CRYSTAL PHYSICS

A TEXT-BOOK
ON
CRYSTAL PHYSICS

by

W. A. WOOSTER, M.A., Ph.D.

Lecturer in the Department of Mineralogy and Petrology
University of Cambridge

CAMBRIDGE
AT THE UNIVERSITY PRESS
1938

CAMBRIDGE
UNIVERSITY PRESS

University Printing House, Cambridge CB2 8BS, United Kingdom

Cambridge University Press is part of the University of Cambridge.

It furthers the University's mission by disseminating knowledge in the pursuit of education, learning and research at the highest international levels of excellence.

www.cambridge.org
Information on this title: www.cambridge.org/9781316611920

© Cambridge University Press 1938

First published 1938
First paperback edition 2016

A catalogue record for this publication is available from the British Library

ISBN 978-1-316-61192-0 Paperback

DEDICATED TO

PROFESSOR A. HUTCHINSON, F.R.S.
EMERITUS PROFESSOR OF MINERALOGY
UNIVERSITY OF CAMBRIDGE

PREFACE

This book is intended to be a text-book for students at universities. For the proper understanding of the contents a knowledge of the elements of physics, mathematics and crystallography is necessary. A two-fold object has been kept in view—to present the classical treatment of the physical properties of crystals in terms of tensor notation and also to indicate the lines of development of modern theoretical and experimental research. Though many references to original papers have been given no attempt has been made to compile an exhaustive list, because the book is primarily a text-book, not a work of reference.

During the preparation of the material for the book I was enjoying the hospitality of the laboratories of Professor Hassel in Oslo, Professor Siegbahn in Upsala, and Professor Wasastjerna in Helsingfors. To them and the University Library Authorities in these places I am much indebted. I wish also to acknowledge the valuable suggestions and criticisms of portions of the text which I have received from Professor P. P. Ewald, Professor K. S. Krishnan, Dr A. J. P. Martin, Professor N. F. Mott, F.R.S., Professor R. Peierls, Dr F. Coles Phillips and Professor C. E. Tilley; and the invaluable help of all kinds which I have received from my wife during the whole of the writing of the book.

Professor A. Hutchinson, to whom this book is dedicated, died while it was in proof, and I should like to record here my appreciation of the stimulus he gave to the study of Crystal Physics during his tenure of the Chair of Mineralogy in Cambridge.

W. A. WOOSTER

January 1938

CONTENTS

INTRODUCTION

The history of the development of crystal physics is a good example of how Science develops in response to a social need. Until recently crystals were only important commercially as jewels; hence only the properties which enabled them to be distinguished from one another were of social interest—a fact which probably explains why crystal optics was almost fully worked out by the middle of last century, although the study of the other physical properties of crystals was only beginning. The thermodynamics of physical processes in crystals and the mathematical formulation of the relations between the physical constants and the crystal symmetry were established during the second half of the nineteenth century and the first decade of the present one. Relatively little experimental work was done during this time, and in particular there was an almost complete absence of researches on the variation of a given physical property in a large number of crystals having chemical or crystallographic similarities. Researches were mostly on one or two well-crystallised substances, and designed to test the mathematical formulation rather than to find out facts about crystals. In surveying the study of crystal physics one of the most striking features is the paucity of experimental data; with few exceptions the physical constants of crystals have only been completely determined for rock-salt, calcite, quartz and Rochelle salt. In 1912 a discovery was made which profoundly affected the study of crystal physics; this was the discovery of the diffraction of X-rays by crystals. From then onwards the arrangements of atoms in crystals began to be studied, and gradually the relations between the physical properties and the crystal structure emerged. A much greater insight into the nature of crystals was thereby obtained, and not only was it sometimes possible to deduce the type of structure from the physical properties but also to invent new crystals with particular physical characters.

The development of crystal physics during the last thirty years has been very uneven, and this again illustrates the impact of industry on scientific research. Among the subjects which in recent years have figured most prominently in the publications on crystal physics are single metal crystals, insulators, and dielectrics, rectification and photoconduction in crystals and piezo-electricity. The methods of growing single crystals of metals are technically so simple that they could have been used a century earlier. These discoveries, in fact, came at a time when industry was making great demands for metals with special properties and when the interest in the physical properties of crystals had been stimulated by the discovery of their structures. Problems connected with the world-wide transmission of electric power at high voltages have necessitated an intensive research on insulators. Dielectric properties of crystals have been studied just when the technique of radio transmission demands millions of condensers. The discovery of the rectifying properties of certain crystals has profoundly changed certain aspects of the technical applications of electricity, and this has greatly stimulated the further study of these properties. The photoconductivity of alkali-halides has been intensively studied during recent years, by means of a technique which could have been used much earlier. The present researches coincide with the rapid development of the photographic and film industry. The application of piezo-electric oscillations has introduced extremely accurate timekeepers, a means of surveying submarine landscapes, of detecting underwater obstructions and submarines, and further provides excellent gramophone pick-ups.

The branches of crystal physics which have been neglected are those with little industrial application. Thus although the theory of the thermal conductivity was worked out about a century ago very few measurements of thermal conductivities in crystals have been made. The characteristic phenomena of rotatory polarisation were partially studied in a few crystals only, and then no further systematic work was done. The first complete study of an optically active biaxial crystal is still to be made. Quantitative measurements on pyro-electricity were carried out on a few crystals to test the theories which were proposed, but of all the hundreds of such crystals which are known to exist no others have since been investigated.

From being a scientific curiosity crystal physics has become in recent times important to industry. This fact is probably not unconnected with the writing of the present text-book. Not all of the subjects which are of technical importance have been treated here, because some of them, for example, the photoelectric and rectifying properties, are mainly a wealth of experimental facts with no comprehensive theory to link them together. Ferromagnetic crystals have not been dis-cussed, because only a few of them are known and there is as yet no extensive study of the relation between the phenomena of ferromagnetism and the crystalline state. Since no text-book devoted entirely to crystal physics has previously been written in English, care has been taken to express in detail the essential principles of the classical analysis of Voigt. The modern theories of the mechanics of crystal lattices and of electrical and thermal conduction in crystals have not been included, because the mathematical knowledge of the students for whom this book is intended is not equal to dealing with these subjects. One of the fundamental lines of advance must be the correlation of the physical properties with the crystal structures and the constituent atoms or ions comprising the crystals. It therefore seemed worth while to introduce the student to the results even though they are incomplete.

The system according to which crystal systems and classes are here described is that given in Volume I, page 33, of the *International Tables for the Determination of Crystal Structures*

(Borntraeger, Berlin, 1935), though, since much of the older literature is expressed in terms of the notation devised by Schoenflies, this is sometimes also given. Where the source of the data on the physical properties of crystals is not stated it has generally been obtained from the *Physikalisch-chemische Tabellen,* Dritter ergänzungsband, Llandolt-Börnstein (Julius Springer, Berlin, 1936).

Some chapters and paragraphs are distinguished by asterisks. These generally indicate that they are more mathematical in treatment than the others. Though some of these paragraphs are essential to a full understanding of the subject they have been so chosen that they may be omitted at a first reading of the book.

CHAPTER I

THE APPLICATION OF TENSOR NOTATION TO CRYSTAL PHYSICS

1.* The transformation of vectors, and second and higher order tensors.

In order to express the mathematical relations between physical quantities with the fewest symbols, a tensor notation is used throughout this book.† We shall use axes OX_1, OX_2, OX_3 (see fig. 1), which are mutually perpendicular (Cartesian coordinates). A vector can then be represented by a symbol p_i, where it is understood that i has all three values 1, 2, 3, and that p_1 is the length of the component of p along OX_1, p_2 and p_3 are the lengths of the corresponding components along OX_2 and OX_3. One of the most important concepts in the theoretical treatment of the physics of crystals is the *transformation of axes*. The changes of the axes included in the word "transformation" are considered in detail in paragraph 2. During these transformations the axes remain mutually perpendicular. We shall denote quantities after transformation by adding

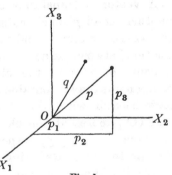

Fig. 1

dashes to the same letters which represented them before the transformation. The new positions of the axes of coordinates are defined by quantities c_{ik}, which are their direction-cosines,‡ i.e. the cosines of the angles between the new axial directions

---

† See H. Jeffreys, *Cartesian Tensors* (Camb. Univ. Press, 1931).
‡ For further reference see R. J. T. Bell, *Coordinate Geometry of Three Dimensions* (Macmillan, London, 1912).

and the old ones. Each new axis has three such c_{ik}'s to define it, and, conversely, each old axis may be defined relative to the new axes by three of these c_{ik}'s. This is expressed by the following scheme:†

	X_1'	X_2'	X_3'
X_1	c_{11}	c_{12}	c_{13}
X_2	c_{21}	c_{22}	c_{23}
X_3	c_{31}	c_{32}	c_{33}

From this, we obtain by reading downwards that OX_1' is defined in relation to OX_1, OX_2 and OX_3 by *direction-cosines* c_{11}, c_{21}, c_{31}, respectively; OX_2 is defined in relation to OX_1', OX_2', OX_3' by the direction-cosines c_{21}, c_{22}, c_{23}, respectively.

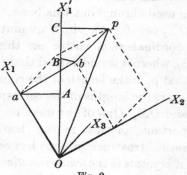

Fig. 2

A vector p transforms to another vector p' on changing the axes of reference, and the relation between the new components of p' and the old components may be obtained as follows. In fig. 2 Op and OX_1' have been drawn in the plane of the paper, and OX_1, OX_2 and OX_3 represent the old axes. Oa, ab, bp represent the components p_1, p_2, p_3, respectively. aA, bB, pC are perpendicular to OX_1'. Referring to the direction-cosine scheme above it will be clear that

$$OA = Oa \cdot c_{11} = p_1 c_{11},$$
$$AB = ab \cdot c_{21} = p_2 c_{21},$$
$$BC = bp \cdot c_{31} = p_3 c_{31}.$$

† This scheme corresponds to that used by H. Jeffreys, *Cartesian Tensors*, and A. J. McConnell, *Applications of the Absolute Differential Calculus* (Blackie, London, 1931), but not to that employed by Th. Liebisch, *Physikalische Krystallographie* (von Veit, Leipzig, 1891).

Now OC is the projection of Op on OX_1' and hence is equal to p_1'.

Since
$$OC = OA + AB + BC,$$
$$p_1' = c_{11}p_1 + c_{21}p_2 + c_{31}p_3.$$

Similarly, by projecting Op on OX_2' and OX_3', we may show that

$$p_2' = c_{12}p_1 + c_{22}p_2 + c_{32}p_3,$$
$$p_3' = c_{13}p_1 + c_{23}p_2 + c_{33}p_3.$$

If instead of projecting the old components of p on a new axis we had projected the new components on each of the old axes in turn, we should have obtained the old components expressed in terms of the new ones. The results are:

$$p_1 = c_{11}p_1' + c_{12}p_2' + c_{13}p_3',$$
$$p_2 = c_{21}p_1' + c_{22}p_2' + c_{23}p_3',$$
$$p_3 = c_{31}p_1' + c_{32}p_2' + c_{33}p_3'.$$

These results may be written in tensor notation as follows:

$$p_i' = c_{ki}p_k,$$
$$p_i = c_{ik}p_k'.$$

It will be noticed that an unrepeated suffix in the equation represents only one of the numbers 1, 2 or 3, but a repeated suffix implies that there are as many terms as can be obtained by substituting 1, 2 and 3 in turn for the repeated suffix and that all of them are to be added together.

Second order tensors.

We shall often have occasion in the following chapters to deal with relations between two vector quantities (p_i, q_k), of the form
$$p_i = a_{ki}q_k.$$

This equation implies that each component of p is a linear function of all the components of q. The nine quantities a_{ki} form a tensor of second order, and each of these nine quantities is called a component of the tensor. The relation between each a_{ki}' after transformation of the axes and all the a_{ki}'s before transformation is of great importance for our study. The formal proof of the relation is given in Appendix I and only the result is quoted here. To make the meaning clear a particular example is given in full. Thus

$$a'_{13} = c_{11}c_{13}a_{11} + c_{11}c_{23}a_{12} + c_{11}c_{33}a_{13}$$
$$+ c_{21}c_{13}a_{21} + c_{21}c_{23}a_{22} + c_{21}c_{33}a_{23}$$
$$+ c_{31}c_{13}a_{31} + c_{31}c_{23}a_{32} + c_{31}c_{33}a_{33},$$

or, in tensor notation,

$$a'_{13} = c_{l1}c_{m3}a_{lm}.$$

In general the relation reads

$$a'_{ik} = c_{li}c_{mk}a_{lm}. \qquad \ldots\ldots(1)$$

Thus a second order tensor may be defined as a set of nine quantities each of which changes according to equation (1) when the axes of reference are changed.

Each tensor component referred to the old axes may be expressed in terms of all the components referred to the new axes by the equation

$$a_{ik} = c_{il}c_{km}a'_{lm}.$$

Third and higher order tensors.

A relation between a second order tensor t_{ik} and a vector s_l is often given by a relation of the form

$$t_{ik} = a_{iik}s_l.$$

The twenty-seven quantities a_{lik} form a tensor of third order. It is shown in Appendix I that the transformation of this tensor leads to the relations

$$a'_{lik} = c_{ml}c_{ni}c_{ok}a_{mno},$$
$$a_{lik} = c_{lm}c_{in}c_{ko}a'_{mno}.$$

The generalisation of this result to a tensor of any order follows at once according to the equations

$$a'_{pqr\ldots w} = c_{Pp}c_{Qq}c_{Rr}\cdots c_{Ww}a_{PQR\ldots W},$$
$$a_{pqr\ldots w} = c_{pP}c_{qQ}c_{rR}\cdots c_{wW}a'_{PQR\ldots W}.$$

2.* The operations of symmetry.

The symmetry operations which are relevant in the study of the physical properties of crystals are those of centre, plane, and diad, triad, tetrad and hexad axes. The operation of a centre of symmetry is such that all corners, edges and faces are inverted through the origin, or otherwise expressed, every face has one parallel to it at the same distance from the origin. When a plane of symmetry is present every corner, edge and face has a corresponding corner, edge or face bearing a mirror-image relation to it, the mirror coinciding with the plane of symmetry. Axes of symmetry usually operate in such a way that after a rotation of $2\pi/n$, where $n = 2, 3, 4$ or 6 (the axis being diad, triad, tetrad or hexad, respectively), the crystal comes into a congruent position. There are also inversion axes of triad, tetrad and hexad symmetry. These operate in such a way that after a rotation through $2\pi/n$ the edges, corners and faces must be inverted through the origin to bring them into congruence with the original position of the crystal.

Certain combinations of symmetry elements are equivalent to others. Thus a diad axis perpendicular to a plane of symmetry necessitates the presence of a centre of symmetry, and this combination is therefore equivalent to a centre plus a plane of symmetry. The same result as that given by a hexad axis may be obtained by superimposing a diad and a triad axis.

When the physical properties are centro-symmetrical (see p. 12), then even in those classes where a polar axis is present the coefficients of the physical property along the polar axis are the same in either direction. This means that there is no distinction between the crystal classes which possess and those which are without a centre of symmetry. Table I shows the reduction of the 32 classes to 11 by this operation.

There is a further reduction in the number of classes which need be considered because the limitations on the coefficients

$a_{ikl...}$, etc. are the same in all classes of symmetry having axes of higher order than the order of the tensor. Thus the number of a_{ik}'s in the trigonal, tetragonal and hexagonal classes of

TABLE I.† *The division of the 32 crystal classes into non-centro-symmetrical and centro-symmetrical groups*

Crystal system	Triclinic	Monoclinic	Rhombic	Trigonal	
Non-centro-symmetrical classes	1	2, m	mm 222	3	$3m$, 32
Classes obtained when centre is introduced	$\bar{1}$	$2/m$	mmm	$\bar{3}$	$\bar{3}m$

Crystal system	Tetragonal		Hexagonal		Cubic	
Non-centro-symmetrical classes	4, $\bar{4}$	42, $4mm$ $\bar{4}2m$	6, $\bar{6}$	62, $6mm$ $\bar{6}2m$	23	43, $\bar{4}3m$
Classes obtained when centre is introduced	$4/m$	$4/mmm$	$6/m$	$6/mmm$	$m3$	$m3m$

symmetry is the same. This reduces the 11 classes mentioned above to 5 for a second order tensor. When the tensor is of the third order the limitations imposed by the tetrad and hexad axes are the same and so on.

3.* The transformation of axes of reference corresponding to the thirty-two classes of crystallographic symmetry.

The general method of studying the limitation of the number of the a_{ik}'s by the symmetry is to transform the axes of reference. The two orientations are related by the symmetry elements. The a_{ik}'s obtained with the new orientation of the axes must be the same as the corresponding a_{ik}'s of the old orientation, since the physical properties, like the faces, are unaffected by a change of orientation corresponding to the

† The symbols for the crystal classes are taken from the *International Tables for the Determination of Crystal Structures*, vol. I, p. 33 (Borntraeger, Berlin, 1935). See also W. H. Bragg and W. L. Bragg, *The Crystalline State*, p. 86 (Bell, London, 1933).

symmetry. The orientations of the axes X_1, X_2, X_3† after transformations by the simpler elements of symmetry are as follows:

a centre gives $\qquad -X_1, -X_2, -X_3,$

a plane perpendicular to X_3 gives

$$X_1, X_2, -X_3,$$

a diad axis parallel to X_3 gives

$$-X_1, -X_2, X_3.$$

Applying this to the five symmetry groups which must be distinguished in dealing with second order centro-symmetrical tensors, we obtain the transformations of Table II.

It will be seen that the uniaxial group is subdivided; the first subdivision does not contain a horizontal diad axis (nor the equivalent vertical plane of symmetry). The horizontal diad axis in the holo-symmetric subdivision is taken parallel to X_1, but the result remains the same whichever horizontal axis is chosen.

4.* The limitations imposed on the number of components of a second order tensor by the requirements of crystal symmetry.

Triclinic system.

It has already been mentioned that the physical properties characterised by second order tensors which we shall consider are centro-symmetrical. This is implicit in any linear equation of the form

$$p_i = a_{ki}q_k,$$

for if we substitute $-p_i$ for p_i and $-q_k$ for q_k, i.e. reverse the directions of the physical vector quantities, the equations are still satisfied if the a_{ik}'s retain their previous values. It is worth noting that if it should be proved experimentally that the relation between p_i and q_k is not the same on reversing

† The axes are often denoted X_1, X_2, X_3 instead of OX_1, OX_2, OX_3.

them, then the present theory would not be true.[†] There are therefore no limitations imposed on the a_{ik}'s by the symmetry, and all nine can be non-zero.

TABLE II. *The non-zero components of a second order tensor in the seven crystal systems*

Symmetry group	Orientation of axes X_1, X_2, X_3 after transformation	Scheme of direction-cosines of original and transformed axes	Non-zero a_{ik}'s
I	$-X_1, -X_2, -X_3$	$\begin{array}{rrr} -1 & 0 & 0 \\ 0 & -1 & 0 \\ 0 & 0 & -1 \end{array}$	$\begin{array}{ccc} a_{11} & a_{12} & a_{13} \\ a_{21} & a_{22} & a_{23} \\ a_{31} & a_{32} & a_{33} \end{array}$
$2/m$	$-X_1, -X_2, X_3$ (diad axis parallel to X_3)	$\begin{array}{rrr} -1 & 0 & 0 \\ 0 & -1 & 0 \\ 0 & 0 & 1 \end{array}$	$\begin{array}{ccc} a_{11} & a_{12} & 0 \\ a_{21} & a_{22} & 0 \\ 0 & 0 & a_{33} \end{array}$
mmm	$-X_1, -X_2, X_3$	$\begin{array}{rrr} -1 & 0 & 0 \\ 0 & -1 & 0 \\ 0 & 0 & 1 \end{array}$	
	$-X_1, X_2, -X_3$	$\begin{array}{rrr} -1 & 0 & 0 \\ 0 & 1 & 0 \\ 0 & 0 & -1 \end{array}$	$\left.\begin{array}{ccc} a_{11} & 0 & 0 \\ 0 & a_{22} & 0 \\ 0 & 0 & a_{33} \end{array}\right\}$
	$X_1, -X_2, -X_3$	$\begin{array}{rrr} 1 & 0 & 0 \\ 0 & -1 & 0 \\ 0 & 0 & -1 \end{array}$	
$\bar{3}, 4/m,$[‡] $6/m$	$X_2, -X_1, X_3$	$\begin{array}{rrr} 0 & -1 & 0 \\ 1 & 0 & 0 \\ 0 & 0 & 1 \end{array}$	$\left.\begin{array}{ccc} a_{11} & a_{12} & 0 \\ -a_{12} & a_{11} & 0 \\ 0 & 0 & a_{33} \end{array}\right\}$
	$-X_1, -X_2, X_3$	$\begin{array}{rrr} -1 & 0 & 0 \\ 0 & -1 & 0 \\ 0 & 0 & 1 \end{array}$	
$\bar{3}m, 4/mmm,$[‡] $6/mmm$	Same as for $\bar{3}$, etc., plus		
	$X_1, -X_2, -X_3$	$\begin{array}{rrr} 1 & 0 & 0 \\ 0 & -1 & 0 \\ 0 & 0 & -1 \end{array}$	$\begin{array}{ccc} a_{11} & 0 & 0 \\ 0 & a_{11} & 0 \\ 0 & 0 & a_{33} \end{array}$
$m3, m3m$	Same as for rhombic system, plus		
	X_2, X_3, X_1	$\begin{array}{rrr} 0 & 0 & 1 \\ 1 & 0 & 0 \\ 0 & 1 & 0 \end{array}$	$\left.\begin{array}{ccc} a_{11} & 0 & 0 \\ 0 & a_{11} & 0 \\ 0 & 0 & a_{11} \end{array}\right\}$
	X_3, X_1, X_2	$\begin{array}{rrr} 0 & 1 & 0 \\ 0 & 0 & 1 \\ 1 & 0 & 0 \end{array}$	

[‡] Tetragonal transformations are the easiest to express analytically.

[†] It was once believed that the thermal conductivity of tourmaline was greater when heat flowed towards the analogous pole than when it flowed away from it, but the experiments were not confirmed. S. P. Thompson and O. J. Lodge, *Phil. Mag.* (5), **8**, 18 (1879); Fr. Stenger, *Ann. Phys.* **22**, 522 (1884).

Monoclinic system.

From the direction-cosines in the transformations given in the previous paragraph we can easily determine which a_{ik}'s are equal to zero. Consider a'_{31} as a particular example.

$$a'_{31} = c_{13}c_{11}a_{11} + c_{13}c_{21}a_{12} + c_{13}c_{31}a_{13}$$
$$+ c_{23}c_{11}a_{21} + c_{23}c_{21}a_{22} + c_{23}c_{31}a_{23}$$
$$+ c_{33}c_{11}a_{31} + c_{33}c_{21}a_{32} + c_{33}c_{31}a_{33}.$$

Only the direction-cosines of c_{11}, c_{22}, c_{33} are different from zero (see Table II), and hence

$$a'_{31} = c_{33}c_{11}a_{31} = 1 . -1 . a_{31} = -a_{31}.$$

Since the transformation corresponds to a symmetry operation $a'_{31} = a_{31}$ and a_{31} can only equal $-a_{31}$ if it is zero. Similarly, we may show

$$a_{32} = a_{23} = a_{13} = a_{31} = 0.$$

But $$a'_{12} = c_{11}c_{22}a_{12} = a_{12},$$

hence a_{12} is not equal to zero, since it is unaltered by the transformation.

Similarly, we find that all the remaining a_{ik}'s have values not equal to zero. Thus for the monoclinic system the coefficients a_{ik} are

$$\begin{matrix} a_{11} & a_{12} & 0 \\ a_{21} & a_{22} & 0 \\ 0 & 0 & a_{33} \end{matrix}$$

Rhombic system.

In the rhombic system limitations are introduced first by a diad axis parallel to X_3 (as in the monoclinic system), then by a diad axis parallel to X_2 and lastly by one parallel to X_1. The effect of the axis parallel to X_3 is to eliminate all a_{ik}'s having a single 3 and, similarly, the other axes eliminate all a_{ik}'s having a single 2 or 1 respectively. Hence we can write down the non-zero a_{ik}'s corresponding to these three transformations thus:

$$\begin{array}{ccc} a_{11} & a_{12} & 0 \\ a_{21} & a_{22} & 0 \\ 0 & 0 & a_{33} \end{array} \qquad \begin{array}{ccc} a_{11} & 0 & a_{13} \\ 0 & a_{22} & 0 \\ a_{31} & 0 & a_{33} \end{array} \qquad \begin{array}{ccc} a_{11} & 0 & 0 \\ 0 & a_{22} & a_{23} \\ 0 & a_{32} & a_{33} \end{array}$$

The only a_{ik}'s common to all three results are a_{11}, a_{22} and a_{33}, which are therefore the only non-zero values.

Tetragonal system.

For a rotation of 90° about the X_3 axis the direction-cosine scheme is

$$\begin{array}{ccc} 0 & -1 & 0 \\ 1 & 0 & 0 \\ 0 & 0 & 1 \end{array}$$

from which it follows that

$$a'_{12} = c_{21}c_{12}a_{21} = 1 \cdot -1 \cdot a_{21} = -a_{21} = a_{12},$$
$$a'_{22} = c_{12}c_{12}a_{11} = -1 \cdot -1 \cdot a_{11} = a_{11}.$$

The a_{ik}'s in this system must also satisfy the relation that they shall remain unchanged on rotating the crystal through 180°, i.e. the same condition as for the monoclinic system. All a_{ik}'s with a single 3 accordingly vanish. Thus the a_{ik} scheme is

$$\begin{array}{ccc} a_{11} & a_{12} & 0 \\ -a_{12} & a_{11} & 0 \\ 0 & 0 & a_{33} \end{array}$$

In those classes of this system which possess a horizontal diad axis parallel to X_1 all a_{ik}'s having a single 1 are eliminated. Thus the only a_{ik}'s which remain are a_{11} and a_{33}.

Cubic system.

The cubic transformations always correspond to a cyclical interchange of the axes X_1, X_2 and X_3 and therefore add to those of the rhombic system the requirement that

$$a_{11} = a_{22} = a_{33}.$$

Hence there is only one a_{ik}.

These results are summarised in the last column of Table II.

Trigonal and hexagonal systems.

The symmetry operations for triad and hexad axes correspond to rotations about the X_3 axis of 120° and 60°, respectively. We can consider them both at once by putting the angle of rotation of the axes $= \theta$, where $\theta \neq 90°$ or 180° (fig. 3). The axes then transform according to the following scheme:†

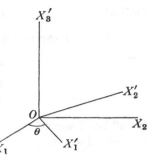

Fig. 3. Diagram showing the rotation of the axes of reference through an angle θ about the axis X_3 (which coincides with X'_3).

$$\begin{array}{cccc} & X'_1 & X'_2 & X'_3 \\ X_1 & c & -s & 0 \\ X_2 & s & c & 0 \\ X_3 & 0 & 0 & 1 \end{array}$$

Then it follows that

$$a'_{11} = c_{11}c_{11}a_{11} + c_{11}c_{21}a_{12} + c_{21}c_{11}a_{21} + c_{21}c_{21}a_{22}$$
$$= c^2 a_{11} + cs a_{12} + cs a_{21} + s^2 a_{22} = a_{11};$$

hence
$$-s^2(a_{11} - a_{22}) + cs(a_{12} + a_{21}) = 0. \quad \ldots\ldots(2)$$

$$a'_{12} = c_{11}c_{12}a_{11} + c_{11}c_{22}a_{12} + c_{21}c_{12}a_{21} + c_{21}c_{22}a_{22}$$
$$= -cs a_{11} + c^2 a_{12} - s^2 a_{21} + cs a_{22} = a_{12}$$
$$-cs(a_{11} - a_{22}) - s^2(a_{12} + a_{21}) = 0. \quad \ldots\ldots(3)$$

Solving equations (2) and (3) in terms of $(a_{11} - a_{22})$ and $(a_{12} + a_{21})$ we find, since $s \neq 0$, that

$$a_{11} - a_{22} = a_{12} + a_{21} = 0;$$

hence
$$a_{11} = a_{22},$$
$$a_{12} = -a_{21}.$$

Further,
$$a'_{13} = c_{11}c_{33}a_{13} + c_{21}c_{33}a_{23}$$
$$= c a_{13} + s a_{23} = a_{13}$$

and
$$(c-1)a_{13} + s a_{23} = 0; \quad \ldots\ldots(4)$$

$$a'_{23} = c_{12}c_{33}a_{13} + c_{22}c_{33}a_{23}$$
$$= -s a_{13} + c a_{23} = a_{23}$$

and
$$-s a_{13} + (c-1)a_{23} = 0. \quad \ldots\ldots(5)$$

† c and s are abbreviations for $\cos \theta$ and $\sin \theta$ respectively.

Solving equations (4) and (5) we obtain

$$a_{13} = a_{23} = 0;$$

similarly $$a_{31} = a_{32} = 0.$$

Hence the non-vanishing a_{ik}'s are the same for the two systems, viz.

$$
\begin{matrix}
a_{11} & a_{12} & 0 \\
-a_{12} & a_{11} & 0 \\
0 & 0 & a_{33}
\end{matrix}
$$

It will be observed that this is the same as the scheme for the tetragonal system.

5. The physical properties of crystals represented by triaxial ellipsoids.

The physical properties to be considered here are set out in the first column of Table III.† Each physical property is defined by coefficients, the values of which are the ratios of the numerical expressions of the two vector quantities given in the second and third columns.

TABLE III. *Physical properties represented by second order tensors*

Physical property	Vector quantities related in the coefficients	
Thermal conductivity	Heat flow	Temperature gradient
Electrical conductivity	Electric current	Potential gradient
Thermoelectricity	Electric current	Thermal gradient
Diamagnetism and para-magnetism	Induced magnetic moment	Magnetic field strength
Dielectricity	Induced electric moment	Electric field strength

6.* Relation between the coefficient defining a physical property in an arbitrary direction and a triaxial ellipsoid.

In studying the anisotropy of crystals it is necessary to determine how the conductivity, susceptibility, etc. vary with

† Optical properties will not be dealt with at length because other text-books deal fully with this subject.

the direction in which they are measured. This is most easily done by supposing the axes to be transformed so that the new X_3 axis, i.e. X_3', coincides with the arbitrary direction under consideration. (X_1' and X_2' need not be specified.) The value of a_{33}' is then the coefficient relating the components along X_3' of the two vectors defining the physical property.

From equation (1), p. 4, it follows that a_{33}' is related to the a_{ik}'s by the expression

$$a_{33}' = c_{13}^2 a_{11} + c_{23}^2 a_{22} + c_{33}^2 a_{33} + c_{13}c_{23}(a_{12} + a_{21})$$
$$+ c_{13}c_{33}(a_{13} + a_{31}) + c_{23}c_{33}(a_{23} + a_{32}).$$

It may be shown experimentally (see Appendix II) that

$$a_{12} = a_{21}, \quad a_{13} = a_{31}, \quad a_{23} = a_{32},$$

so that

$$a_{33}' = c_{13}^2 a_{11} + c_{23}^2 a_{22} + c_{33}^2 a_{33} + 2c_{13}c_{23}a_{12} + 2c_{13}c_{33}a_{13} + 2c_{23}c_{33}a_{23}.$$
$$\ldots\ldots(6)$$

This equation is that of a surface of second order, usually a triaxial ellipsoid having a general orientation with respect to the axes of reference. Further, the length of the radius vector r is equal to

$$\frac{1}{\sqrt{a_{33}'}}.$$

7.* Orientation of the triaxial ellipsoid representing a_{33}' relative to the crystallographic axes in the crystal systems.

Applying the limitations on the non-zero a_{ik}'s derived in paragraph 4 to the equation for a_{33}' we obtain the values of a_{33}' set out in Table IV.

These equations are the analytical expression of the geometrical relation which exists between the axes of the triaxial ellipsoid and the crystallographic axes. The least symmetry the ellipsoid can possess is three mutually perpendicular diad axes. If the crystal has no symmetry then the relation between the axes of the ellipsoid and the crystallographic axes must be quite general, corresponding to the first equation in Table IV.

If the crystal has one diad axis then this must coincide with one of the principal axes of the ellipsoid, for only with such a

TABLE IV. *The variation with direction of the component a'_{33} of a second order tensor in the seven crystal systems*

Crystal system	a'_{33} is equal to
Triclinic	$c_{13}{}^2 a_{11} + c_{23}{}^2 a_{22} + c_{33}{}^2 a_{33} + 2c_{13}c_{23}a_{12} + 2c_{13}c_{33}a_{13} + 2c_{23}c_{33}a_{23}$
Monoclinic	$c_{13}{}^2 a_{11} + c_{23}{}^2 a_{22} + c_{33}{}^2 a_{33} + 2c_{13}c_{23}a_{12}$
Rhombic	$c_{13}{}^2 a_{11} + c_{23}{}^2 a_{22} + c_{33}{}^2 a_{33}$
Trigonal ⎫ Tetragonal ⎬ Hexagonal ⎭	$c_{13}{}^2 a_{11} + c_{23}{}^2 a_{11} + c_{33}{}^2 a_{33}$
Cubic	$(c_{13}{}^2 + c_{23}{}^2 + c_{33}{}^2)\, a_{11} = a_{11}$

mutual orientation would the ellipsoid conform to the crystal symmetry. In the same way in the rhombic system all three principal axes of the ellipsoid must coincide with those of the crystal and we obtain the equation

$$c_{13}^2 a_{11} + c_{23}^2 a_{22} + c_{33}^2 a_{33} = a'_{33},$$

which corresponds to the usual expression for an ellipsoid, namely,

$$\frac{x^2}{a^2} + \frac{y^2}{b^2} + \frac{z^2}{c^2} = 1.$$

In the uniaxial crystal systems the ellipsoid must have equal axes perpendicular to the principal crystallographic axis, and hence it becomes a surface of revolution. Finally, in the cubic system the ellipsoid degenerates into a sphere.

CHAPTER II

HOMOGENEOUS DEFORMATION

THERMAL EXPANSION AND
PLASTIC DEFORMATION

1. Introduction

In this chapter we shall consider the thermal expansion and the homogeneous deformation of crystals. These properties are taken together because of the geometrical similarity in the result produced by change of temperature, and that produced by permanent deformation under homogeneous shear. A crystal also suffers homogeneous deformation when subjected to hydrostatic pressure, but consideration of this is deferred to Chapter VIII. A deformation is said to be homogeneous when all lines which were straight before the deformation are also straight afterwards. Apart from the formal relationship there is little in common between thermal expansion and plastic deformation.

Thermal expansion was one of the first branches of crystal physics to be studied. The extremely careful observation by Mitscherlich on the change of angle with temperature between neighbouring faces of a calcite rhomb led to the early study of the anisotropy of thermal expansion. The interferometric work associated with Fizeau led to a great increase in the accuracy of measurement. Although the pioneer work was done so long ago, very little has as yet been done towards a comprehensive survey of thermal expansion, even in those substances which can be obtained as large crystals. X-ray methods of measuring thermal expansions have recently been developed and, as large crystals are not required, the field of study has been greatly extended. X-ray diffraction effects depend simply on the size, nature and distribution of atoms within a unit cell. Although very many of these cells are required to produce a measurable intensity of reflection, the macroscopic size of the crystal does not affect the determination of the atomic positions and thus the expansion may be measured when only extremely small crystals are available. Perhaps the most important application of the measurements of thermal expansion is to the study of transitions from one crystalline form to another; fundamental changes in the crystal structure are usually accompanied by discontinuities or sudden changes in the rate of expansion.

The plastic deformation of crystals has evoked an enormous amount of inquiry. The reason for this is not far to seek. Our knowledge of the properties and uses of metals has increased very much during the last twenty years. There is an ever-growing demand for metals and alloys with particular properties. In single-crystal form many metals exhibit quite remarkable plasticity. An ordinary polycrystalline wire of cadmium which cannot be stretched appreciably by a tension of several kilograms becomes so ductile after being converted into a single crystal that it bends under its own weight. Although it has assumed so much importance in the last few years, the study of plastic deformation is not new. The early work was carried out on ionic crystals such as rock-salt and calcite and revealed most of the laws relating to the processes involved. The number of studies of the mechanical properties of rock-salt is very great, and the recent intensive study has been directed to obtaining a clearer idea of the processes of plastic flow.

2.1.* General theory.

When a body is subjected to homogeneous deformation any three points which were on a straight line before the deformation are on another straight line after it. Further, lines which were parallel before the deformation remain parallel after, and lines which were coplanar remain coplanar. A circle becomes an ellipse or, more generally, a sphere becomes a triaxial ellipsoid.

The simplest analytical expression of a homogeneous deformation is that the components (u_1, u_2, u_3) of the displacement of a point having coordinates (x_1, x_2, x_3) are given by

$$u_1 = r_{11}x_1 + r_{21}x_2 + r_{31}x_3 \quad \text{etc.,}$$

or $\qquad\qquad u_i = r_{ki}x_k.$ $\qquad\qquad\qquad$ ......(1)

We shall show that this equation gives rise to the changes characteristic of homogeneous deformation.

Consider a straight line passing through two points

$$(x_1, x_2, x_3) \quad \text{and} \quad (x_1 + dx_1, x_2 + dx_2, x_3 + dx_3)$$

for which the displacements are

$$(u_1, u_2, u_3) \quad \text{and} \quad (u_1 + du_1, u_2 + du_2, u_3 + du_3),$$

respectively.

It is clear that if both points received the same displacement $(du_i = 0)$ the line joining them would be straight and parallel to the original line. In fact, the displacement of the second point is a little different from the first, and we have to find how it varies with the distance between them. Since each component of the displacement may be an independent function of each of the coordinates, we may write

$$du_1 = \frac{\partial u_1}{\partial x_1}dx_1 + \frac{\partial u_1}{\partial x_2}dx_2 + \frac{\partial u_1}{\partial x_3}dx_3;$$

but from equation (1)

$$\frac{\partial u_1}{\partial x_1} = r_{11}, \quad \text{etc.} \quad \text{or} \quad \frac{\partial u_i}{\partial x_k} = r_{ki},$$

therefore $\qquad\qquad du_i = r_{ki}dx_k.$

The difference between the displacement of the second point and the first is thus a linear function of the components of the line joining them. This can only be true if the displaced line is straight. A two-dimensional figure may help to illustrate this.

Two points, A (x_1, x_2) and B $(x_1 + dx_1, x_2 + dx_2)$, are indicated in fig. 4. A is displaced to D $(AC = u_1, CD = u_2)$ and B to H $(BE + FG = u_1 + du_1, EF + GH = u_2 + du_2)$. The differences in the x_1 and x_2 coordinates of A and B are dx_1 and dx_2, respectively. Our proof above shows that

$$du_1 = r_{11}dx_1 + r_{21}dx_2, \quad du_2 = r_{12}dx_1 + r_{22}dx_2.$$

Thus, if the distance between A and B is doubled along the same line, the components du_1 and du_2 are doubled also. Hence the direction of the line through D and H is the same

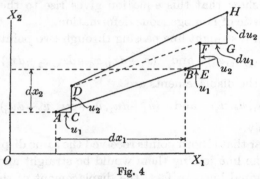

Fig. 4

no matter what the distance of H from D. Further, since du_1, du_2 depend only on dx_1, dx_2 and not on the actual values of (x_1, x_2), the angle HDF would be the same wherever D was in the body. Hence all lines originally parallel to AB will be parallel to DH after the deformation.

A direct consequence of this is that a rectangle deforms into a parallelogram, a circle into an ellipse, and a sphere into a triaxial ellipsoid. An important special case arises if in place of equation (1) we write

$$\left.\begin{aligned} u_1 &= r_{11}x_1 \\ u_2 &= r_{22}x_2 \\ u_3 &= r_{33}x_3 \end{aligned}\right\}. \qquad \ldots\ldots(2)$$

Consider any point (x_1, x_2, x_3) on the surface of a sphere of unit radius. After the deformation its coordinates have become (x_1^*, x_2^*, x_3^*), where

$$\begin{aligned} x_1^* &= x_1 + u_1 = x_1(1 + r_{11}), \\ x_2^* &= x_2 + u_2 = x_2(1 + r_{22}), \\ x_3^* &= x_3 + u_3 = x_3(1 + r_{33}). \end{aligned}$$

Further, $\qquad x_1^2 + x_2^2 + x_3^2 = 1,$

and therefore

$$\frac{x_1^{*2}}{(1+r_{11})^2} + \frac{x_2^{*2}}{(1+r_{22})^2} + \frac{x_3^{*2}}{(1+r_{33})^2} = 1. \qquad \ldots\ldots(3)$$

This is the equation to an ellipsoid the semi-axes of which are of lengths $(1+r_{11})$, $(1+r_{22})$, $(1+r_{33})$ and coincide with the axes of reference.

Thus when the axes of reference are chosen so as to coincide with the principal axes of the ellipsoid arising through the deformation, the simpler expression (2) for the components of deformation may be used instead of the more general form (1).

2.2.* Components of strain.

We fix upon any given point in the homogeneously deformed body and take this as our origin. The displacement under the stress of any other point having coordinates (x_1, x_2, x_3) is taken to be (u_1, u_2, u_3). Then (as shown on p. 17) the meaning of homogeneity of deformation may be expressed analytically thus:

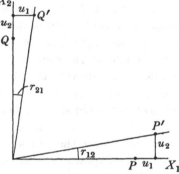

$$u_1 = r_{11}x_1 + r_{21}x_2 + r_{31}x_3 \text{ etc.,}$$

or, in general,

$$u_k = r_{1k}x_1 + r_{2k}x_2 + r_{3k}x_3.$$

The quantities r_{ik} are called the components of strain and transform according to the usual relation for second order tensors.

Fig. 5. Diagram showing the rotation of the axes X_1, X_2 corresponding to the components of strain r_{12} and r_{21}.

If we differentiate the equation for u_1 we obtain

$$\frac{\partial u_1}{\partial x_1} = r_{11},$$

so that the extension per unit length in the X_1 direction is r_{11}. The quantities r_{ik}, where $i \neq k$, are closely related to the angles of rotation of the lines which before the deformation were parallel to the axes of reference. For point P (fig. 5), lying on the OX_1 axis, $x_2 = x_3 = 0$; hence

$$u_1 = r_{11}x_1,$$
$$u_2 = r_{12}x_1.$$

Hence the rotation of the OX_1 axis

$$= \frac{u_2}{x_1} = r_{12}.$$

Similarly, for the point Q, $x_1 = x_3 = 0$, and hence

$$u_1 = r_{21}x_2, \quad u_2 = r_{22}x_2$$

and the rotation of the OX_2 axis

$$= \frac{u_2}{x_2} = r_{21}.$$

If there is to be no net rotation as a result of the deformation $r_{12} = r_{21}$. Actually the displaced positions of P and Q will not lie in the plane X_1OX_2, so that the condition $r_{12} = r_{21}$ will be satisfied, provided the projections of the new OX_1, OX_2 axes on the plane X_1OX_2 make equal angles with OX_1 and OX_2.

If the axes of reference were mutually perpendicular before the deformation and have become triclinic as a result of the deformation, it is always possible to arrange the new axes relative to the old so that there is no resultant rotation, i.e. so that if the deformed axes be projected on to the old axial planes the displacement of the projections from their original positions are equal and opposite.† Hence we may write

$$r_{ik} = r_{ki}.$$

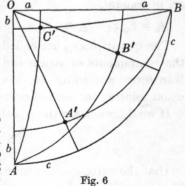

Fig. 6

† This may be seen as follows: Suppose OAB to be one quadrant of a stereographic projection representing the direction of the axes X_1, X_2, X_3 before the deformation. Draw two zones making angles a with AO and AB respectively, as shown in fig. 6. Draw two further zones making an angle b with BO and BA. The point C' now represents the deformed

3.1. The application of the theory of homogeneous deformation to the thermal expansion of crystals.

The changes in shape and size of a crystal on changing its temperature are completely defined by the directions of the principal axes of the ellipsoid referred to in paragraph 2 relative to the crystallographic axes and by the values of the principal coefficients of expansion α_{11}, α_{22} and α_{33}. It is often convenient to represent the expansion of the crystal in a given direction in terms of the changes in length of the corresponding radius vector. If unit vector in an arbitrary direction changes to a vector of $1+g$ after the deformation we may obtain the value of g from equation (3). Neglecting second powers of the small quantities α_{11}, α_{22}, α_{33}, this may be written:

$$x_1^{*2}(1-2\alpha_{11})+x_2^{*2}(1-2\alpha_{22})+x_3^{*2}(1-2\alpha_{33}) = 1,$$

$$x_1^{*2}+x_2^{*2}+x_3^{*2}-1 = 2(\alpha_{11}x_1^{*2}+\alpha_{22}x_2^{*2}+\alpha_{33}x_3^{*2}).$$

Now $\qquad\qquad x_1^{*2}+x_2^{*2}+x_3^{*2} = (1+g)^2 = 1+2g.$

Hence $\qquad\qquad g = \alpha_{11}x_1^{*2}+\alpha_{22}x_2^{*2}+\alpha_{33}x_3^{*2};$

since $x_1^{*2} = x_1^2(1+2\alpha_{11})$ we may, if we neglect squares of α_{11}, etc., write

$$g = \alpha_{11}x_1^2+\alpha_{22}x_2^2+\alpha_{33}x_3^2.$$

Since the vector joining the point (x_1, x_2, x_3) to the origin is of unit length, x_1, x_2 and x_3 are the direction-cosines of this vector. *It is important to remember that throughout this book an arbitrary direction is denoted by the number 3 and therefore these direction-cosines are usually referred to as* c_{13}, c_{23}, c_{33}, *meaning by* c_{13} *the cosine of the angle between the X_1 axis and the new direction 3, and so on.* Thus the last equation may be written:

$$g = \alpha_{11}c_{13}^2+\alpha_{22}c_{23}^2+\alpha_{33}c_{33}^2. \qquad\qquad \ldots\ldots(4)$$

position of the X_3 axis. Draw two further zones making an angle c with OA and OB. The points A', B' define the orientation of the new X_1, X_2 axes. It will be seen that the new set of axes satisfies the conditions mentioned above, that the projection in pairs on the original axial planes makes equal angles with the corresponding pairs of original axes. Further, the triangle $A'B'C'$ is completely general and may be made to have any form by altering the angles a, b, c.

This is an equation to a surface of second degree; if α_{11}, α_{22}, α_{33} are all positive it represents a triaxial ellipsoid, but if one or two of them are negative this method of representation requires some modification, since we cannot represent in a single three-dimensional figure both expansion and contraction.

3.2. Lines of zero expansion.

Calcite affords an excellent example of the simultaneous occurrence of positive and negative coefficients. α_{11}, which is equal to α_{22}, since the crystal is uniaxial, is negative, while α_{33} is positive. If we put $g = 0$ to correspond to zero expansion, we have, from equation (4),

$$0 = \alpha_{11}(c_{13}^2 + c_{23}^2) + \alpha_{33}c_{33}^2;$$

but since

$$c_{13}^2 + c_{23}^2 + c_{33}^2 = 1,$$

$$0 = \alpha_{11}(1 - c_{33}^2) + \alpha_{33}c_{33}^2$$

$$= \alpha_{11} + c_{33}^2(\alpha_{33} - \alpha_{11}).$$

Hence

$$c_{33}^2 = \frac{\alpha_{11}}{\alpha_{11} - \alpha_{33}}.$$

If θ is the angle between the arbitrary direction under consideration and the trigonal axis (X_3)

$$c_{33} = \cos\theta.$$

Hence

$$\tan^2\theta = \frac{-\alpha_{33}}{\alpha_{11}}.$$

For calcite $\alpha_{33} = 24 \cdot 91 \times 10^{-6}$; $\alpha_{11} = -5 \cdot 56 \times 10^{-6}$, hence $\theta = 64° 43'$.

Thus the directions which remain of the same length are all inclined at the same angle to the trigonal axis and form a cone of semi-angle $64° 43'$.

4. Validity of the law of rational indices, of the anharmonic ratio relation, etc. in an expanded crystal.

We have seen in paragraph (2.1) that every plane face remains plane after the crystal has experienced a homogeneous deformation and that all parallel lines remain parallel and expand to the same extent. Thus the intercept which a plane cuts off on one of the crystallographic axes and the axis itself expand to the same extent, and the reciprocal of their ratio, which is equal to one of the face indices, remains integral and constant. Similarly, since the coordinates of any point in the crystal referred to the crystallographic axes bear the same ratio to the axial lengths before and after deformation, the zone axis symbols all remain constant. Hence, all the crystallographic relations between face and zone indices remain unchanged.

5. Experimental methods of determining principal thermal coefficients of expansion.

5.1. By combination of change of interfacial angle and bulk expansion.

The principle of this method may be illustrated by an application to the thermal expansion of calcite.[†] Referred to hexagonal axes the crystal expands along the c axis and contracts along all directions perpendicular to this. From the resulting change of angle between the rhombohedral faces the change in axial ratio may be calculated. If $\alpha_{\parallel}(\alpha_{33})$ and $\alpha_{\perp}(\alpha_{11})$ are the principal thermal expansions and $(c/a)_1$ and $(c/a)_0$ are the axial ratios at temperatures t_1 and t_0, then

$$\left(\frac{c}{a}\right)_1 = \left(\frac{c}{a}\right)_0 \frac{1+\alpha_{\parallel}(t_1-t_0)}{1+\alpha_{\perp}(t_1-t_0)}.$$

Then, neglecting $\alpha_{\perp}(t_1-t_0)$ in comparison with unity,

$$\frac{(c/a)_1-(c/a)_0}{(c/a)_0} = (\alpha_{\parallel}-\alpha_{\perp})(t_1-t_0).$$

From these measurements $(\alpha_{\parallel}-\alpha_{\perp})$ is determined. The volume of a unit cube of this material after the same change in temperature is

$$(1+\alpha_{\parallel})(1+\alpha_{\perp})(1+\alpha_{\perp}) = 1+\alpha_{\parallel}+2\alpha_{\perp}.$$

If the bulk coefficient of expansion is τ,

$$\tau = \alpha_{\parallel}+2\alpha_{\perp}.$$

† The detailed application of this method to all non-cubic crystals is given in Th. Liebisch, *Physikalische Krystallographie*, p. 71 (von Veit, Leipzig, 1891).

By combining the values of $(\alpha_\parallel + 2\alpha_\perp)$ and $(\alpha_\parallel - \alpha_\perp)$ we can find $\alpha_\parallel$ and $\alpha_\perp$ separately.

5.2. By measurement of change of thickness of crystal blocks.

The apparatus devised originally by Fizeau and later improved by Tutton is essentially as follows. The crystal block A is placed on a tripod B which carries a plane glass plate, C, on the screws which form the feet. Light reflected from the upper surface of the crystal interferes with the light reflected from

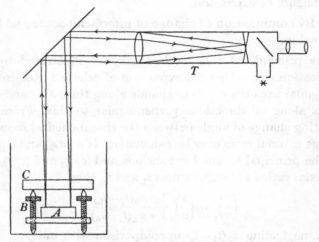

Fig. 7. Diagram of the apparatus used in the interferometric method of measuring thermal expansion.

the lower surface of the glass plate. The interference fringes are observed in the telescope T. The light is generally introduced by a Gaussian eyepiece. The plate C is a wedge of small angle, and therefore the light reflected from its upper face does not interfere with light from the lower surface. The glass plate can be arranged to give either circular or straight fringes, but whichever are used the difference in path introduced by moving the lower plate so that each fringe moves into the position previously occupied by its neighbour is half a wave-length. The stand carrying the crystal expands simultaneously with it, and hence the change in the distance between the crystal and the glass plate depends on the material and length of the legs. Tutton† has utilised the large expansion of aluminium to compensate for the expansion of the platinum-iridium legs by placing the crystal on a block of aluminium of suitable size. Any change in the position of the fringes is thus caused by the crystal alone. High accuracy is obtainable with this method, but its application is limited to substances of which adequate crystal plates may be obtained.

5.3. Number of measurements necessary in the various crystal systems.

In the following paragraph the measurements which must be made in the various crystal systems to determine completely the constants of expansion are discussed in detail. The same considerations apply to any second order tensor.

Anorthic system.

When the principal axes of expansion have been found three crystal plates perpendicular to these axes suffice to determine the principal coefficients of expansion. The determination of the principal axes is often a tedious process, because it is necessary to proceed by trial and error. The results may be checked by employing plates of other known orientations.

Monoclinic system.

Most crystals belong to the monoclinic system, which is therefore of special importance. Since one principal axis must coincide with the diad axis of the crystal only two axes lying in the (010) plane remain to be found. For this purpose

---

† A full description of the experimental method is given in Tutton's *Crystallography and Practical Crystal Measurement*, p. 1308 (Macmillan and Co., London, 1922).

measurements of the expansion in three directions in the plane (010) are required. The relative positions of the crystallographic axes a, c, the principal expansion axes, which are the axes of reference, OX_1, OX_3, and the arbitrary direction lying in the plane (010) in which the expansion has been measured (denoted OX_3') are shown in fig. 8. Then if α_{11}, α_{33}, α_{33}' are the expansion coefficients along X_1, X_3 and X_3', respectively,

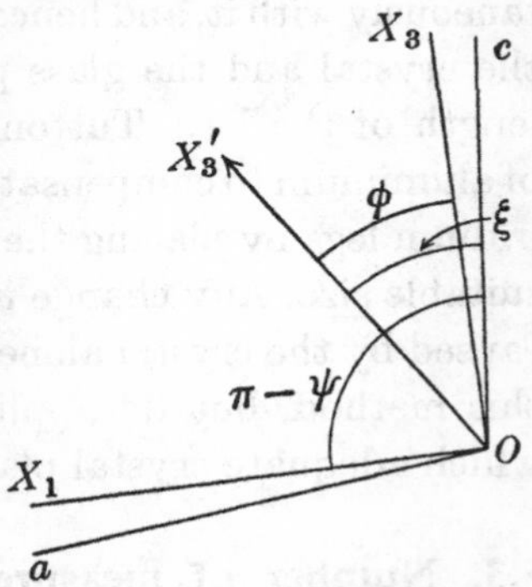

$$\alpha_{33}' = c_{13}^2 \alpha_{11} + c_{33}^2 \alpha_{33},$$

or, if we insert the functions of the angles shown in fig. 8,

Fig. 8. Diagram showing the meaning of ψ in a monoclinic crystal.

$$\alpha_{33}' = \sin^2 \phi \, \alpha_{11} + \cos^2 \phi \, \alpha_{33}$$
$$= \tfrac{1}{2}(1 - \cos 2\phi)\,\alpha_{11} + \tfrac{1}{2}(1 + \cos 2\phi)\,\alpha_{33}$$
$$= \frac{\alpha_{11} + \alpha_{33}}{2} - \frac{\alpha_{11} - \alpha_{33}}{2} \cos 2\phi.$$

Since $\qquad \pi - \psi\dagger - \dfrac{\pi}{2} = \xi - \phi, \qquad \phi = -\dfrac{\pi}{2} + (\xi + \psi),$

and

$$\alpha_{33}' = \frac{\alpha_{11} + \alpha_{33}}{2} + \frac{\alpha_{11} - \alpha_{33}}{2}\,(\cos 2\xi \cos 2\psi - \sin 2\xi \sin 2\psi).$$

Now put

$$\left.\begin{array}{c} \dfrac{\alpha_{11} + \alpha_{33}}{2} = A \\[2mm] \dfrac{\alpha_{11} - \alpha_{33}}{2} \cos 2\psi = B \\[2mm] -\dfrac{\alpha_{11} - \alpha_{33}}{2} \sin 2\psi = C \end{array}\right\} . \qquad \dots\dots(5)$$

Then $\qquad\qquad \alpha_{33}' = A + B \cos 2\xi + C \sin 2\xi.$

† ψ is usually defined as the acute angle between OX_1 (the direction of the greatest expansion coefficient in the plane (010)) and the crystallographic c-axis. ψ is positive when OX_1 lies between $+a$ and $+c$ and negative when it lies between $-a$ and $+c$ (as in fig. 8).

Since α'_{33} and ξ are measurable, three sets of measurements give three simultaneous equations from which A, B and C may be found. From the relations (5), we have

$$-\tan 2\psi = \frac{C}{B},$$

$$\alpha_{11} = A + \frac{B}{\cos 2\psi},$$

$$\alpha_{33} = A - \frac{B}{\cos 2\psi}.$$

Thus all the required information may be obtained.

Rhombic system.

Three plates cut perpendicular to the crystallographic axes afford directly the principal coefficients of expansion.

Hexagonal, rhombohedral and tetragonal systems.

Two plates only are necessary, and for preference these should be cut perpendicular and parallel to the principal axis. Plates of general orientation can also be used, but their expansion is a function of both the principal coefficients of expansion.

Cubic system.

All orientations of the plate give the same result.

5.4. By means of X-rays.

The arrangement of the atoms of a crystal on a lattice makes it possible to mark out families of identical, equally spaced, parallel planes of atoms. Such families of parallel planes reflect X-rays according to Bragg's law, which is expressed by the equation
$$\lambda = 2d \sin \theta,$$

where λ is the wave-length of the X-rays, d the spacing of the planes and θ the glancing angle of reflection. (It should be noted that d is different according to the order of reflection, so that if the actual distance between the planes is D and the order of reflection n, then $d = D/n$.) The experimental measurement consists in finding a change in the angle θ caused by the change of temperature. Differentiating the above expression, we have

$$0 = d\cos\theta \,.\, \delta\theta + \delta d\sin\theta,$$

or

$$\delta d = -d\,.\,\cot\theta\,.\,\delta\theta$$

$$= -\frac{\lambda}{2}\operatorname{cosec}\theta\,.\,\cot\theta\,.\,\delta\theta. \qquad \ldots\ldots(6)$$

From this expression it follows that $\delta\theta$ is large when θ approaches $90°$ for a given value of δd. The accuracy of measurement depends simply on the ease with which this angle $\delta\theta$ may be found, and hence it is generally necessary to conduct the experiment in such a way that the glancing angle of reflection is nearly $90°$, i.e. the reflected rays are almost returning along their original path. A camera constructed to admit of measurements of θ close to $90°$ is employed for this purpose. The usual form consists of a cylindrical box on the inside of which the film is fixed and the powder or crystal is mounted on the axis of the box. A hole is cut in the film to admit the slits, and the required reflections are arranged to be symmetrically placed with respect to this hole. When the developed film is laid flat their separation may be accurately measured. Considerable practical difficulties arise in making θ sufficiently near to $90°$. The only quantity in equation (6) which can be independently varied is λ, and hence it is necessary to employ an X-ray tube with a variety of anticathodes, if full advantage is to be taken of the method. Sometimes it is possible to vary the choice of d, but usually the number of planes which can be identified with complete certainty is limited to a very few planes with low indices.

The particular details which arise in applying this method to various crystal systems are given below.

Anorthic system.

Each crystal needs to be treated on its own merits, and although simplifications are sometimes possible no general simple analysis can be given.

Monoclinic system.

Theoretically the three principal coefficients of expansion and their orientation with respect to the crystallographic axes may be obtained from changes in spacing of four planes none of which is parallel to any other. In practice it is found simplest to use three planes of the $(h0l)$ type and one $(0k0)$ plane. The results of the measurements on the three $(h0l)$ planes are used in the way described on p. 26.

Rhombic system.

In this system the changes in spacing or changes in angle of glancing reflection, θ, of three general planes suffice to give the principal coefficients of thermal expansion. It is, of course, also necessary to know the cell size. The spacing of a general plane is related to the indices by the formula

$$\frac{h^2}{a^2} + \frac{k^2}{b^2} + \frac{l^2}{c^2} = \frac{1}{d_{hkl}^2} = \frac{4\sin^2\theta}{\lambda^2},$$

and hence on differentiating

$$-2\left(\frac{h^2}{a^3}\partial a + \frac{k^2}{b^3}\partial b + \frac{l^2}{c^3}\partial c\right) = \frac{8\sin\theta\,\cos\theta\,\partial\theta}{\lambda^2},$$

$$\frac{h^2}{a^3}\partial a + \frac{k^2}{b^3}\partial b + \frac{l^2}{c^3}\partial c = \frac{-2\sin 2\theta \cdot \partial\theta}{\lambda^2}. \qquad \ldots\ldots(7)$$

The values of all quantities in equation (7) are found from the experiments, except ∂a, ∂b, ∂c; hence if $\partial\theta$ be determined from three planes of different indices ∂a, ∂b, ∂c may be calculated. The principal coefficients of expansion may then be found from the relation

$$\alpha_1 = \frac{\partial a}{a\,\partial t}, \text{ etc.,}$$

where ∂t is the temperature change corresponding to $\partial\theta$.

Hexagonal and rhombohedral systems.

These two systems may be taken together if the same system of indexing the planes be applied to both. The relation between the spacing and indices is

$$\frac{4}{3}\frac{(h^2+hk+k^2)}{a^2}+\frac{l^2}{c^2}=\frac{1}{d_{hkl}^2}=\frac{4\sin^2\theta}{\lambda^2},$$

and hence

$$\frac{4}{3}\frac{(h^2+hk+k^2)}{a^3}\,\partial a+\frac{l^2}{c^3}\,\partial c=\frac{-2\sin 2\theta}{\lambda^2}\,\partial\theta.$$

Observations on two planes are enough to determine the principal coefficients $\alpha_{\parallel}$ and $\alpha_{\perp}$.

Tetragonal system.

The formulae in this system are as follows:

$$\frac{h^2+k^2}{a^2}+\frac{l^2}{c^2}=\frac{1}{d_{hkl}^2}=\frac{4\sin^2\theta}{\lambda^2},$$

and hence

$$\frac{(h^2+k^2)}{a^3}\,\partial a+\frac{l^2}{c^3}\,\partial c=\frac{-2\sin 2\theta}{\lambda^2}\,\partial\theta.$$

Here also observations on two planes suffice to fix $\alpha_{\parallel}$ and $\alpha_{\perp}$.

Cubic system.

Putting $a=c$ and $\partial a=\partial c$ in the previous equation, we obtain

$$\frac{h^2+k^2+l^2}{a^3}\,\partial a=\frac{-2\sin 2\theta}{\lambda^2}\,\partial\theta\quad\text{or}\quad\alpha=-\cot\theta.\frac{\partial\theta}{\partial t}.$$

This result shows that any plane is equally suitable for determining the expansion of a cubic crystal.

6. The variation of the thermal expansion with temperature—the detection of transformations.

Researches on LiF, NH_4Cl, calcite, aragonite and sphalerite[†] at temperatures down to $-250°$ C. have shown that in accordance with Nernst's theorem the thermal expansion

† H. Adenstedt, *Ann. Phys.* **26**, 69 (1936).

coefficients tend to vanish as the absolute zero is approached. Thus the values at room temperature (R.T.) and at the lowest temperatures used (L.T.) for the thermal expansions of the substances mentioned above are given in the following table:

TABLE V. *Comparison of the thermal expansions of certain substances at room temperature and at about* $-250°$ C.

Substance	Expansion coefficients ($\times 10^6$)					
	R.T.	L.T.	R.T.	L.T.	R.T.	L.T.
LiF	31	2				
NH$_4$Cl	27	6				
CaCO$_3$ (calcite)	-6	-0.9	—	—	25	0.6
CaCO$_3$ (aragonite)	10	3	16	1	33	6
ZnS (sphalerite)	6	-0.1				

If the variation of expansion with temperature be compared with the corresponding change of the specific heat at constant pressure it is in general true that they are in a constant ratio in conformity with Grüneisen's law.

Other researches on zinc,[†] beryllium,[‡] mercury,[§] beryllium oxide,[||] zinc oxide[||] have also shown increases in the thermal expansion coefficients as the temperature rises. This is by no means invariably found, however, and there are many remarkable exceptions. Thus the expansion coefficient of NH$_4$Cl rises rapidly to a maximum at $-31°$ C. and falls equally rapidly on either side of that temperature. The expansion coefficient at $-31°$ C. is some thirty times greater than the values at $-30°$ and $-32°$ C. Ni[¶] has an expansion coefficient of 14 at $0°$ C. and this rises sharply above $300°$ C., so that at $370°$ C. it is 26. At $390°$ C. it has fallen again to 15 and thereafter rises slowly to 18 at $600°$ C. These are examples of a

† E. A. Owen and E. L. Yates, *Phil. Mag.* 17, 113 (1934).
‡ E. A. Owen and T. L. Richards, *Phil. Mag.* 22, 304 (1936); G. F. Kossolapow and A. K. Trapesnikow, *Z. Krist.* 94, 53 (1936).
§ D. M. Hill, *Phys. Rev.* 48, 620 (1935).
|| H. Braekken and O. Jore, *Norges Tekniske Høiskole.* Avhandlinger til 25 års Jubileet (1935), p. 75.
¶ E. A. Owen and E. L. Yates, *Phil. Mag.* 21, 809 (1936).

change in the magnitude of the thermal expansion coefficient which does not correspond to any known change in the crystal structure. Other notable examples of this phenomenon are afforded by manganous oxide, ferrous oxide, manganese sulphide and magnetite, Fe_3O_4,[†] at temperatures of about $-100°$ C. When the crystal structure changes, either because certain groups, e.g. NO_3, NH_4, start to rotate, instead of simply oscillating about their mean position, or because certain atoms change their parameters, as in the change from α- to β-quartz at $573°$ C., there is usually a corresponding abrupt change in the expansion coefficients.

7. Relation between the coefficients of thermal expansion and the nature of the atoms or ions composing the crystal.

The variation of the coefficients of expansion of the elements with atomic number is roughly periodic, as may be seen from Table VI. The largest coefficients of expansion occur among

TABLE VI. *Thermal expansion coefficients of the elements*
(× 10⁶)

Li 56	Be 11 14	B 8	C 1				
Na 71	Mg 27 25	Al 23	Si 4		S 67 78 20		
K 83 Cu 17	Ca 28 Zn 55 14	Ga 18	Ge 8	As 29 3	Cr 8	Mn 22 Br 67	Fe 12, Co 12, Ni 14
Rb 90 Ag 19	Cd 49 17	In 45 12	Sn 46 26	Nb 7 Sb 17 8	Mo 5 Te −2 27	I 73	Ru 7, Rh 9, Pd 12
Cs 97 Au 14	Hg 50 38	Tl 72 9	Pb 29	Bi 17 12	W 7	Re 12 5	Os 7, Ir 7, Pt 9

(Where two numbers follow a chemical symbol the upper one refers to the expansion coefficient parallel to the principal axis and the lower to that perpendicular to this axis.)

† B. S. Ellefson and N. W. Taylor, *J. Chem. Phys.* **2**, 58 (1934).

the halogens and the alkali metals. The irregularities in the periodic variation are numerous and show that other factors than the atomic number are important in determining the thermal expansion.

The thermal expansion coefficients of some of the alkali halides are given in Table VII.

TABLE VII.   *Thermal expansion coefficients of alkali halides*

($\times 10^6$)

	F	Cl	Br	I
Li	34	44	50	59
Na	36	40	43	48
K	37	38	40	45
Rb	—	36	38	43
Cs	32	50	60	55

There is a steady variation of the expansion coefficients with increasing size of anion, though this change is small compared with the difference between the expansion coefficients of, say, sodium chloride and magnesium oxide, which both have the same structure type. Among the various attempts to relate the thermal expansion coefficients to other physical properties the most successful is that which connects expansion with 'electrostatic share'. The 'electrostatic share', q, of an ion in a crystal lattice is defined as the ratio of the valency of the ion to the number of ions of the opposite sign, which are its immediate neighbours. The relation between the thermal expansion, α, and q which has been proposed† is given by the equation

$$\alpha q^2 \simeq 10^{-6}.$$

That this rule holds true for many simple ionic crystals may be seen from Table VIII, but it is not universally applicable:

There are certain crystals for which $\alpha q^2 \times 10^6$ is very far from unity, e.g. PbS for which it is equal to 2·2; rutile, brookite and anatase, the three polymorphic modifications of TiO_2, for which it is equal to 3·6, 2·8 and 8·3, respectively. The rule

† H. D. Megaw, Ph.D. Thesis, Cambridge, 1935.

can be extended to crystals which have more than one sort of anion by attaching 'weights' to the electrostatic share of the

TABLE VIII. *Showing the relation between thermal expansion and 'electrostatic share'*

Crystal	Valency of cation	Number of anion neighbours around cation	q	$\alpha \times 10^6$	$\alpha q^2 \times 10^6$
CsCl	1	8	1/8	50	0·8
NaCl	1	6	1/6	40	1·1
CaF$_2$	2	8	1/4	19	1·2
BeO	2	4	1/2	5·1	1·3
				5·3	1·3
MgO	2	6	1/3	13	1·4
SnO$_2$	4	6	2/3	3·9	1·7
				3·2	1·4

(The expansion coefficients refer to room temperature.)

constituent ions. The expansion coefficients of crystals which are not ionic often fail to conform to this rule; crystals with weak binding forces have large expansion coefficients and values of $\alpha q^2 \times 10^6$ much greater than unity.

8. Relation between the thermal expansion coefficients and the crystal structure of non–cubic crystals.

At present it is not possible to predict the expansion coefficients of a crystal of a given chemical composition and crystal structure. We can, however, group crystals according to the general nature of their structures and find empirically what are the corresponding physical properties. Structures may conveniently be grouped into isosthenic lattices, layer lattices, anti-layer lattices, chain lattices and three-dimensional frameworks. Isosthenic lattices consist of atoms or ions (including symmetrical groups such as SO_4, SiO_4) linked to all their neighbours with bonds of the same strength. Cubic and hexagonal close-packing are typical isosthenic lattices. Layer lattices may be regarded as arising from isosthenic lattices when they are pulled out along one axis—giving rise to a stratified structure in which the binding between the atoms in the layers is stronger than that between the layers. A similar type of structure arises where planar groups such as CO_3, NO_3 are arranged parallel to one another. Anti-layer lattices arise when an isosthenic lattice is compressed along one axis, the binding between atoms parallel to the unique axis being stronger than that between atoms in planes perpendicular to it. Chain lattices are characterised by the strongest bonds between neighbouring atoms forming parallel chains or ribbons running through the structure. Examples of this type of structure are afforded by cinnabar, HgS, and selenium. Structures containing linear ions such as N—N—N arranged parallel to one another often have physical properties similar to those of chain lattices. Three-dimensional frameworks include a great many silicate structures in which SiO_4 and AlO_4 tetrahedra are linked together to form a framework which is not limited to a chain or band but extends so as to include the whole crystal. The other cations or molecular groups fill in the interstices in these frameworks, but considerable changes may usually be made in this filling material without affecting the main framework. Examples of such lattices are quartz, the felspars, and some zeolites. Examples of the expansion coefficients of crystals belonging to these four groups are given below:

TABLE IX. *Thermal expansion of some isosthenic lattices*

Crystal	Structure type†	Expansion coefficients ($\times 10^6$)			Remarks
		$\alpha_\parallel$ or α_a	α_b	$\alpha_\perp$ or α_c	
Mg	A3	27	—	25	Structure contains practically close-packed oxygen ions
BeO.Al$_2$O$_3$	H	6·0	6·0	5·2	
(chrysoberyl)					
Al(F, OH)$_2$SiO$_4$	S-O$_5$	4·8	4·1	5·9	(As for BeO.Al$_2$O$_3$)
(topaz)					
K$_2$SO$_4$	H-1$_6$	36	32	36	
Rb$_2$SO$_4$	H-1$_6$	36	32	35	

† The classification of structure types is that given in the *Strukturbericht*, P. P. Ewald and C. Hermann (Akademische Verlagsgesellschaft, Leipzig, 1931).

It will be observed that the anisotropy in thermal expansion coefficients for crystals belonging to this group is small irrespective of the number of different kinds of atoms they contain.

TABLE X. *Thermal expansion coefficients of crystals having a layer lattice type of structure*

Crystal	Struc-ture type[†]	Expansion coefficients ($\times 10^6$)		Remarks
		$\alpha_{\parallel}$	$\alpha_{\perp}$	
Cd	A3	49	17	$c/a = 1\cdot89$
Zn	A3	55	14	$c/a = 1\cdot86$
As	A7	$\alpha(40°$ to trig. axis) 28	3	Layers parallel to (0001)
Sb	A7	17	8	,, ,,
Bi	A7	17	12	,, ,,
Ca(OH)$_2$	C6	33	10	Layers parallel to (0001)
Mg(OH)$_2$	C6	45	11	,, ,,
Al(OH)$_3$	O7	α_1 38 α_2 11 α_3 $-$ 6 ψ $-46°$		The greatest expansion does not occur in a direction perpendicular to the layers but in an inclined direction
CaSO$_4$.2H$_2$O	H-4$_6$	α_1 29 α_2 42 α_3 2 ψ $-39°$		Layers are perpendicular to α_2 and chains of atoms are nearly parallel to α_3
CaCO$_3$‡	G1	25	-6	Planes of CO$_3$ ions are parallel to (0001)
NaNO$_3$‡	G1	120	11	,, ,,

† The classification of structure types is that given in the *Strukturbericht*, P. P. Ewald and C. Hermann (Akademische Verlagsgesellschaft, Leipzig, 1931).
‡ These crystals are included with the layer lattices because the existence of parallel planes of planar ions gives them many of the physical properties of layer lattices. They are not, however, true layer lattices as is shown by the absence of a unique good cleavage perpendicular to the principal axis.

It will be clear from the above table that in many layer lattices the expansion coefficient perpendicular to the layers is greater than the coefficients corresponding to expansions in the layers. There are two exceptions to this rule provided by bismuth triiodide and thallium; $\alpha_{\parallel}$ and $\alpha_{\perp}$ for bismuth triiodide are respectively 51 and 50, while c/a is 1·60. As the iodine ions are arranged on a hexagonal close-packed pattern but slightly compressed along the trigonal axis, we might expect that $\alpha_{\parallel}$ would be less than $\alpha_{\perp}$. This exception is not important, however, since $c/a = 1·60$ is so near the ideal value for hexagonal close-packing. $\alpha_{\parallel}$ and $\alpha_{\perp}$ for thallium† are 72 and 9, while c/a is 1·60, and in this case also it would be expected that $\alpha_{\parallel}$ should be less than $\alpha_{\perp}$. The expansion of hydragillite, $Al(OH)_3$, shows that a layer lattice can expand not only by increasing the perpendicular distance between the layers but also by a shearing movement of the layers over one another.‡ This is only possible when the symmetry permits it, and hence this is a phenomenon which is restricted to monoclinic and anorthic crystals.

The following table gives the only available examples of the anisotropy of thermal expansion of anti-layer lattices:

TABLE XI. *Thermal expansion coefficients of anti-layer lattices*

Crystal	Structure type	Expansion coefficients ($\times 10^6$)		Remarks
		$\alpha_{\parallel}$	$\alpha_{\perp}$	
Be	A3	11	14	c/a 1·58
ZnO	B4	5·9	6·9	c/a 1·60
BeO	*B4*	*5·1*	*5·3*	*c/a 1·63*

In these two lattices $\alpha_{\parallel}$ is less than $\alpha_{\perp}$, showing that the direction in which the atoms are most compressed is also the direction of least expansion. The expansion coefficients of beryllium oxide, which is an isosthenic lattice, since $c/a = 1·63$, are added for comparison. Although for this crystal $\alpha_{\parallel}$ and $\alpha_{\perp}$ are not equal, they are less different than they are for zinc oxide.

Cinnabar, HgS, is the only example of a chain lattice for which data on the expansion coefficients are available; $\alpha_{\parallel}$ is

† G. Shinoda, *Kyoto Coll. Sci. Mem.* 16, 193 (1933).
‡ H. D. Megaw, *Proc. Roy. Soc.* A, 142, 198 (1933).

equal to 22×10^{-6} and $\alpha_\perp$ to 18×10^{-6}. From the general considerations already developed we should expect that $\alpha_\parallel$ would be less than $\alpha_\perp$ for a chain lattice.

The thermal expansion coefficients of some crystals having three-dimensional framework structures are given below:

TABLE XII. *Thermal expansion coefficients of crystals having three-dimensional framework structures*

Crystal	Structure type	Expansion coefficients ($\times 10^6$)			ψ
		$\alpha_\parallel$ or α_1	α_2	$\alpha_\perp$ or α_3	
SiO_2 (quartz)	C8	9	—	14	
$KAlSi_3O_8$ (adularia)	—	19	−2	−1	−83°
$NaAlSi_3O_8$ (albite)	—	$\alpha_{\parallel a}$ 13	$\alpha_{\perp(010)}$ 4	$\alpha_{\perp(001)}$ 4	
$CaAl_2Si_2O_8$ (anorthite)	—		$\alpha_{\perp(010)}$ 1·6	$\alpha_{\perp(00\ )}$ 6	

It is probably significant that the differences in crystallographic axial lengths between adularia and albite roughly correspond to the magnitudes of the expansion coefficients:

	a	b	c
Adularia	8·45	12·90	7·15
Albite	8·14	12·86	7·17
Difference	0·31	0·04	0·02

The principal thermal expansion axis α_1 is nearly in the direction of a, and we see that the greatest expansion occurs in the same direction both for change of temperature and size of univalent cation.

The felspars are poor examples of three-dimensional framework structures because the folded four-membered rings of AlO_4-SiO_4 parallel to the a axis† make them somewhat similar to chain lattices. Quartz, however, is a good example of the type, and we see that even it has a considerable anisotropy in thermal expansion.

† W. H. Taylor, *Z. Krist.* **85**, 425 (1933), and *Proc. Roy. Soc.* A, **145**, 80 (1934).

9. The application of the theory of homogeneous deformation to the plastic deformation of crystals.

When a solid body is subjected to a stress it is deformed. If it recovers its shape on removal of the stress the deformation is elastic, but if the deformation is permanent then the solid has been plastically deformed. If one observes the behaviour of single crystals under stress by macroscopic methods it is found that a small stress gives rise to elastic deformation, greater stress to plastic flow. In one instance an interferometric method of observation was employed and another phenomenon was found. Single crystals of tin† examined at room temperature were found to deform plastically by gliding even when the stress applied was vanishingly small. The amount of such plastic deformation was limited, and the same effect was produced either by a very small force acting over a long time, or a greater force, up to that corresponding to the elastic limit, over a shorter interval. The rate of deformation was greater, the greater the stress, but whatever the force the final state was approached exponentially.

Plastic deformation can take place in two ways, by gliding and by twinning. Twinning is always a form of homogeneous deformation, but gliding may be non-homogeneous. The deformation by gliding recorded with the tin crystals was homogeneous, but when as a result of the deformation slip bands appear it is inhomogeneous (cf. paragraph 10). Here the gliding occurs on particular lattice planes, parallel to the glide planes, one part of the crystal slipping several thousand cell lengths relative to the other. The planes on which the slipping occurs are more or less equally spaced about 10^4 cell lengths apart. The cause of this regularity is not yet understood.

Although strictly speaking ordinary plastic deformation is not homogeneous, it is macroscopically homogeneous to a degree that enables it to be treated as though it were. The theory of homogeneous deformation has been treated in

† B. Chalmers, *Proc. Roy. Soc.* A, **156**, 427 (1936).

paragraph 2, where it is shown that a sphere marked out in the crystal becomes an ellipsoid on deformation. It is convenient to refer the deformation to the two circular sections of this ellipsoid. Suppose that the plane on which gliding occurs is perpendicular to the paper and the direction of gliding is OA, fig. 9. The amount of slip is defined by a constant s, which is the translation of a plane parallel to OA at unit perpendicular distance from OA. The lines with the arrows indicate the amount of gliding which has taken place at various distances from OA. The two sections which are unchanged in shape and size by deformation are AO and BO. The position

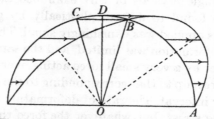

Fig. 9. Diagram showing the deformation of a
circle into an ellipse.

of OB changes as the amount of slip increases, but the position of AO remains constant throughout. The amount of slip is directly related to the angle AOB, since the point B on the ellipse must have been the point C on the original circle, and CB is drawn parallel to OA.

Thus
$$\frac{s}{2 \cdot OD} = \tan BOD = \cot AOB,$$

where OD is the perpendicular from O on BC. A simple construction employing the stereographic projection enables us to find the orientation of any given plane after the deformation, provided the orientation of the unchanged planes be known. Usually a plane of a crystal is represented by a single point in a stereographic projection, but it is equally correct to represent it by a great circle parallel to the plane. This method is adopted in fig. 10.

A plane of arbitrary orientation is represented by the circle VWX and we are required to find its orientation after the deformation. It is clear that the line UX remains unchanged, since it lies in the glide plane PSR, which is one circular section, the other being PQR; before the deformation PQR was PTR, for $TU = UQ$. The point W moves to Y, where $TW = QY$, because WY is a line parallel to the glide direction US, and the plane PQR contains all the points which before

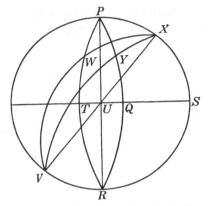

Fig. 10. Diagram showing the construction for finding the orientation of any plane in the crystal after deformation.

the deformation were in the plane PTR. The corresponding position of the plane VWX is therefore given by the great circle VYX. By defining a line as the intersection of two planes we may find, from a similar construction, the corresponding directions of any line in the crystal before and after the deformation.

The geometrical changes introduced by the glide process may therefore be completely specified, either by giving the glide plane, glide direction and amount of glide, or alternately by a statement of the orientation of the circular sections of the ellipsoid.

10. The geometry of the gliding process in various crystals.

Gliding in crystals is restricted to certain crystallographic-ally defined planes and directions. In hexagonal metals, for example zinc and cadmium, the glide plane is (0001) and the possible glide directions are $[2\bar{1}\bar{1}0]$, $[11\bar{2}0]$, $[\bar{1}2\bar{1}0]$; in cubic face-centred metals the glide planes are the four {111} planes and the glide directions any of the twelve [110]. The orientation of the crystal relative to the applied stress determines which of the glide elements come into play, for gliding occurs in the planes and directions suffering the greatest shearing

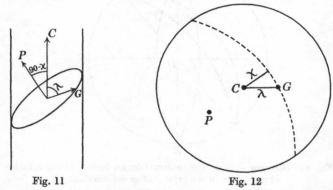

Fig. 11   Fig. 12

Figs. 11, 12. Diagrams showing the orientation of the normal to the glide plane, P, and glide direction, G, with respect to the axis of the rod, C.

stress. Suppose the length of the rod is inclined to the glide direction at an angle λ (fig. 11) and to the normal to the glide plane at an angle $(90 - \chi)$ (assuming there is a single glide plane as in cadmium). In the stereographic projection (fig. 12) the dotted line represents the glide plane, G the glide direction, C the rod direction, and P the normal to the glide plane. If the force per unit area of the rod is Z, then the force acting along the rod per unit area of the section parallel to the glide plane is $Z \sin \chi$. The component of this acting in the glide direction is $Z \sin \chi \cos \lambda$ and hence the shearing stress τ is given by

$$\tau = Z \sin \chi \cos \lambda.$$

P and G are necessarily mutually perpendicular, but C may have any orientation with respect to them. χ can be equal to λ but cannot be greater than this, and the maximum value of $\sin \chi \cos \lambda$ is obtained when $\chi = \lambda = 45°$. In this case $\tau = Z/2$.

Fig. 13 represents the plastic deformation of a cadmium wire. C is the direction of the length of the wire, and G the glide direction. As gliding proceeds C tends to become parallel to the glide direction. This may be shown in a stereographic

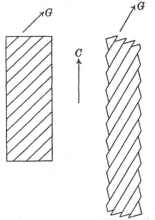

Fig. 13. Diagram illustrating the elongation of a rod produced by gliding.

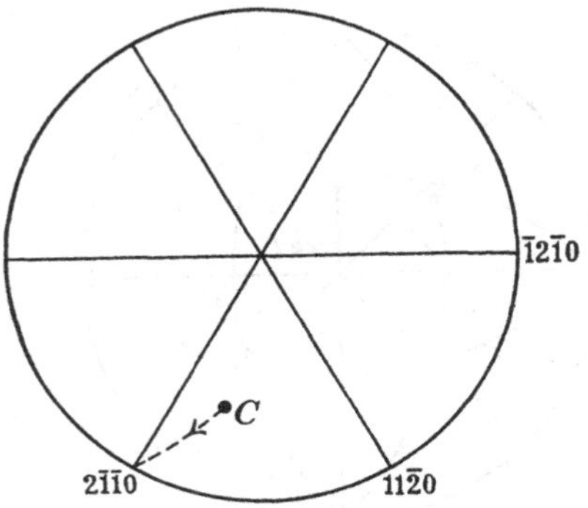

Fig. 14. Diagram showing the change in orientation of the axis of a rod of a single crystal of cadmium when gliding occurs on the basal plane.

projection, fig. 14, where C, the direction of the wire, moves towards the nearest diad axis. Although this simple type of gliding usually occurs when rod-shaped crystals of zinc and cadmium are deformed, complications may arise with certain orientations and at certain temperatures. Reference should be made elsewhere for a description of this work.[†]

A stereographic projection of a cubic face-centred metal is given in fig. 15. Here P is the direction of the rod, the corresponding glide plane is Q and the glide direction is R. It is easy to see by inspection that for no other combination of (111)

† E. Schmid and W. Boas, *Kristallplastizität*, p. 86 (Springer, Berlin, 1935); C. F. Elam, *Distortion of Metal Crystals*, Ch. II. (Oxford Univ. Press, 1935.)

plane and [110] direction is P so nearly midway between them. As the rod is extended P moves towards R, but on crossing the line OS, which is a symmetry plane, the shearing stress becomes greatest for another pair of glide elements, namely U and T. A priori we should expect that the direction of the rod would not cross the symmetry plane but move along the line OS. It is observed, however, that the two glide systems operate unequally and the point representing the length of the rod alternates in the way represented by the

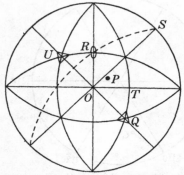

Fig. 15. Stereographic projection showing the favoured glide elements for a given orientation of the axis of the rod relative to the crystallographic axes in a cubic face-centred metal.

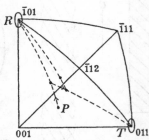

Fig. 16. Portions of a stereographic projection showing the alternations in glide elements which occur for certain orientations of the axis of the rod in cubic face-centred metals.

diagram in fig. 16. When P is in the triangle 001-011-$\bar{1}12$ it moves towards R, and when in the triangle 001-$\bar{1}01$-$\bar{1}12$ it moves towards T, and this continues until it reaches $\bar{1}12$. The direction of the axis remains unchanged relative to the crystallographic axes on further extension.

Other systems of glide planes and glide directions are found in other metals, and the available information is collected in Table XIII.

In ionic substances gliding is essentially the same as in metals. Most of the work done up to the present has been on rock-salt. A peculiarity of the gliding of rock-salt, due to its high symmetry, makes it impossible for gliding to take place

TABLE XIII. *Data on the plastic deformation of metals*

Metal	Degree of chemical purity	Method of preparation	Translation elements		Critical shearing stress (kg./mm.²)	Rigidity modulus (kg./mm.²)	Strain at elastic limit (× 10⁵)
			Glide plane	Glide direction			
Copper	>99·9	Solidified in vacuum	(111)	[10$\bar{1}$]	0·10		
Silver	99·99				0·060		
Gold	99·99				0·092		
Nickel	99·8				0·58		
Magnesium	99·95	Recrystallisation	(0001)	[11$\bar{2}$0]	0·083	1700	4·9$_5$
Zinc	99·96	Drawn out			0·094	4080	2·3
Cadmium	99·996	of the melt			0·058	1730	3·3$_5$
β-Tin	99·99	Drawn out of the melt	(100) (110)	[001]	0·189 0·133	1790 1790	10·6 7·4$_3$
Bismuth	99·9	Drawn out of the melt	(111)	[10$\bar{1}$]	0·221	970	22·8

on one plane only at a time, at least two glide planes being involved simultaneously. This is shown in fig. 17. For the direction P the effective glide elements, i.e. those to which P is most nearly symmetrically disposed, are A (101) and B [$\bar{1}$01], the glide plane and glide direction respectively. The great circle AP cuts the circle 010-$\bar{1}$01 in C and the great circle BP cuts the circle 101-010 in D. Then, in the above nomenclature $PC = \chi$, $PB = \lambda$. If, however, the glide plane is ($\bar{1}$01) and the glide direction [101], we shall denote PD as χ' and PA as λ'.

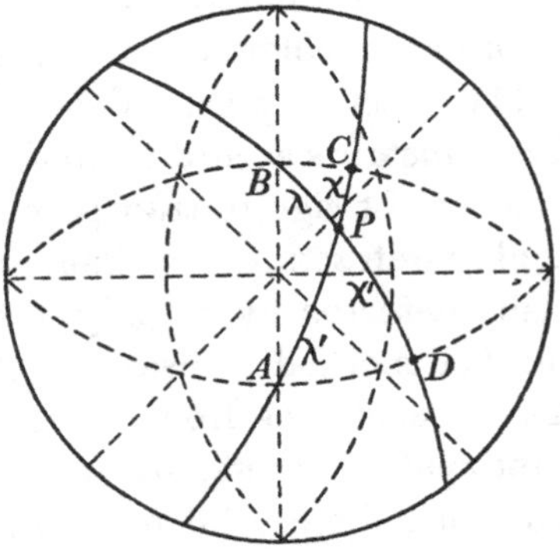

Fig. 17. Stereographic projection showing the equivalence of two glide systems in rock-salt.

The shearing stress using the first glide elements is given by

$$\tau = Z \sin \chi \cos \lambda;$$

now $$\sin\chi = \cos\lambda' \quad \text{and} \quad \cos\lambda = \sin\chi',$$

hence, if the shearing stress using the second set of glide elements is τ',

$$\tau' = Z\sin\chi'\cos\lambda' = Z\sin\chi\cos\lambda = \tau.$$

Hence under all circumstances the stress tending to produce gliding is equally great on both glide planes.

11. Optical methods of studying plastic deformation.

The transparency of ionic crystals has made possible an optical method of investigating plastic flow. The crystal is observed in polarised light between crossed nicols while it is being deformed. The field remains dark until the elastic limit is reached, when a few bright lines appear parallel to glide planes, becoming increasingly numerous as the deformation proceeds (see fig. 18). Another optical method has been employed, making use of crystals coloured by exposure to X-rays or γ-rays or ultra-violet light. Plastic deformation decreases the strength of the colour. On the other hand, if a crystal be irradiated while under a load, the region of plastic deformation colours more quickly than the undistorted parts.

The onset of plastic deformation appears to occur at different instants according to the method used for its detection. The earliest sign in transparent crystals is the appearance of doubly refracting lamellae. For metal crystals it is most easily detected by a study of the stress-strain diagram. Still another method utilises the change in the character of the X-ray reflections from the planes of the crystal, the sharp Laue spots characteristic of the undeformed crystal are drawn out when the crystal becomes deformed, and the amount of general scattering increases.

12. Dynamics of plastic deformation.

A typical stress-strain diagram for a monocrystalline cadmium wire, which shows many of the features associated with gliding, is given in fig. 19. This curve is obtained by

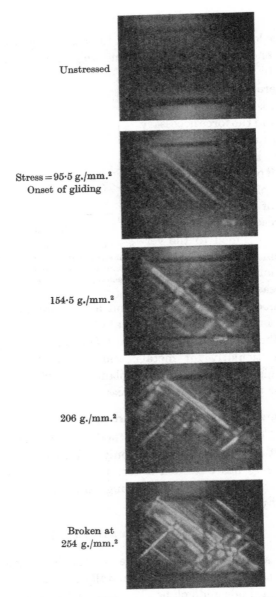

Fig. 18. Photographs of a KCl crystal stretched between crossed nicols. The progressive extension of the crystal is accompanied by a corresponding increase in the number of doubly refracting lamellae parallel to the glide planes. (Reproduced from *Kristallplastizität*, E. Schmid and W. Boas (Springer, Berlin, 1935), by kind permission of the publishers.)

applying a load which is gradually increased, e.g. by fixing the upper end of the wire and adding weights to a scale pan attached to the lower end. A rather different curve is obtained if the wire is stretched at a uniform rate and the force necessary at each instant to continue the stretching is measured. The portion AB of the curve in fig. 19 represents elastic stretching; if the scale of the abscissae were much increased it would be seen that it was inclined to the vertical. From C to D there is a relatively large stretching for a small increase in stress. This is

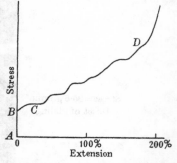

Fig. 19. Diagram illustrating the type of stress-strain curve which is obtained by gradually increasing the load on a single crystal of cadmium.

never a smooth curve, the extension for a small increase in the load being sometimes small and at others great. The stress at which the transition from AB to CD occurs is fairly well defined, particularly for metals, and the tangential force acting at this instant along a glide plane in the glide direction is known as the critical shearing stress.

13. The law of critical shearing stress.

Provided extremely sensitive methods of investigation are not used, it is found that crystals only suffer shear when the shearing stress has exceeded a certain value. This result is known as the law of critical shearing stress and is formally expressed by the equation,

$$Z = \frac{\tau_c}{\sin \chi \cos \lambda} \quad \text{(cf. paragraph 10)},$$

where τ_c is a constant known as the critical shearing stress.

The law has been tested experimentally for various metals which have a single glide plane. The observed values of τ_c are almost independent of χ and λ. Compression normal to the gliding plane during shear gives rise to no change in the value of τ_c. The critical shearing stress depends very much on the rate at which the crystal is grown. The slower the growth the lower is the value of τ_c. Long annealing at a temperature just below the melting point also lowers the shearing stress. τ_c for ionic crystals is less constant than for metals and for both types of crystal it tends to zero as the melting point is reached. However, except near the melting point the critical shearing stress is generally much less sensitive to temperature change than is, for instance, the viscosity of liquids. The addition of impurities usually raises the critical shearing stress, but it does not appear that on extrapolating to absolute purity it would be zero.

14. Plastic deformation by twinning.

In addition to plastic deformation by gliding, deformation also occurs by means of twinning. The classic example of this is found in calcite, which forms artificial twins of great perfection and considerable size. Such twins may be formed by pressing a knife into a polar edge of a calcite rhomb. The shape

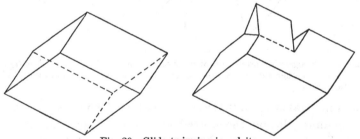

Fig. 20. Glide-twinning in calcite.

of the calcite, before and after, is shown in fig. 20. The crystal glides on the $(01\bar{1}2)$ plane in the $[0\bar{1}11]$ direction. The faces of the twinned part remain optically flat, in sharp distinction to the faces of a metal crystal which has been deformed by gliding. This optical flatness proves that the deformation is homogeneous, i.e. that every lattice plane moves the same amount as every other, relative to its immediate neighbours. The position of the twinned crystal relative to the original is

W C P   4

invariable, the relative movement of two planes parallel to the glide plane and unit distance apart being characteristic of the substance.

Although this type of deformation has always been called 'twinning', it is important to realise that it is not the same as ordinary twinning, i.e. the relation between the faces on the deformed part and the faces of the undeformed crystal is never the same as between the two halves of an ordinary twin. Fig. 21 shows a calcite crystal projected with $(01\bar{1}2)$ vertical

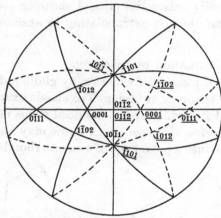

Fig. 21. Stereographic projection of faces of the forms $\{10\bar{1}1\}$, $\{01\bar{1}2\}$, $\{0001\}$ in calcite together with the projection of these same faces in a twin on $(01\bar{1}2)$ (underlined).

and a natural twin on $(01\bar{1}2)$ as twin plane. The faces obtained by twinning lie on the dotted circles and their indices are underlined once. Fig. 22 shows the original crystal and a twin produced by gliding with great circles representing planes (instead of the more usual representations of figs. 21 and 23). The derivation of the artificial twin is shown by the following example. The glide plane is $(01\bar{1}2)$ and the gliding is in the $[0\bar{1}11]$ direction so that the circular sections of the deformation ellipsoid are the planes $(01\bar{1}2)$ and $(\underline{0\bar{1}11})$, represented by the great circles AF and FC respectively. The plane FC was the plane FB before the deformation and this was cut at B by

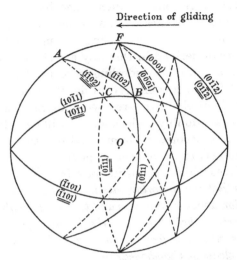

Fig. 22. Stereographic projection of faces of the forms {10Ī1}, {01Ī2}, {0001} in calcite in which each face is represented by a great circle parallel to it. The dotted circles show the positions of the planes after glide-twinning.

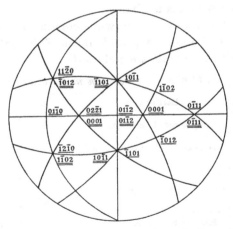

Fig. 23. Stereographic projection of the normals to the faces of the glide-twin (doubly underlined) and a comparison of the indices which these faces have when referred to the axes of the crystallographic twin on (01Ī2) with their original indices.

the plane ($1\bar{1}02$). A remains unchanged by the deformation but B moves to C. Hence the new orientation of the plane ($1\bar{1}02$) is given by the circle passing through A and C. The faces of the twin derived in this way are shown in fig. 22 by dotted circles and doubly underlined indices. The faces of the two kinds of twin are compared in fig. 23, where the indices of the crystallographic twin are once underlined and those of the glide-twin are doubly underlined. It will be noted that the basal plane has become a rhombohedron, two faces of the rhombohedron $\{\bar{1}012\}$ have become prism faces $\{11\bar{2}0\}$ and the only set of faces which retains its form is $\{10\bar{1}1\}$.

It emerges that the important distinction between twinning by simple rotation about a twin axis and twinning by gliding is that all faces preserve their forms in the former type, whereas all except one change their form in the latter. The unique form in calcite is the cleavage rhombohedron.

Twinning by gliding occurs in a number of substances both organic and inorganic.† Twinning of this type is also found in metal crystals. The plane parallel to which the layers glide and the plane which is unchanged in shape and size after the deformation are known as the twin elements, and are denoted K_1 and K_2. In the following table most of the data available about twinning in metals are collected:

TABLE XIV. *Data on metallic twins*

Metal	Lattice type	K_1	K_2
α-Fe	Body centred cubic, A 2	(112)	(11$\bar{2}$)
Be Mg Zn Cd	Hexagonal close-packed, A 3	(10$\bar{1}$2)	(10$\bar{1}\bar{2}$)
β-Sn	Tetragonal, A 5	(331)	(11$\bar{1}$)
As Sb Bi	Rhombohedral, A 7	(011)? (011) (011)	(100) (100) (100)

† Th. Liebisch, *Physikalische Krystallographie*, p. 114 (von Veit, Leipzig, 1891); O. Mügge, *Z. Krist.* **75**, 32 (1930); *ibid.* **76**, 359 (1931).

There is no well-defined force like the critical shearing stress above which twinning occurs in metals. The orientation of the crystal relative to the stress has a profound influence. For example, if the hexagonal axis of a cadmium crystal is nearly

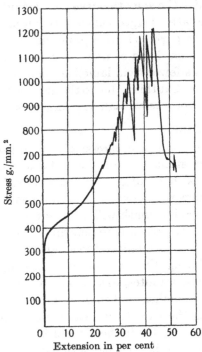

Fig. 24. Stress-strain diagram obtained in an apparatus in which the extension is predetermined and the necessary load measured. The first part of the curve indicates gliding and the second, jagged part, twinning. (Reproduced from *Kristallplastizität*, E. Schmid and W. Boas (Springer, Berlin, 1935), by kind permission of the publishers.)

parallel to the axis of the wire, twinning occurs after very little stretching, but if the hexagonal axis is orientated so that gliding occurs readily, twinning occurs only shortly before fracture. The twinning in cadmium is accompanied by a sharp sound, similar to the 'cry' of tin. When an apparatus is used which measures the force necessary to produce extension at a given rate (instead of the more usual determination of the extension produced by a given force) certain features which distinguish deformation by twinning from deformation by gliding are revealed. The difference between the load-extension curves for the two types of deformation is shown in fig. 24, in which the smooth curve corresponds to the initial gliding and the latter jagged curve to twinning of a cadmium single crystal. Impurities have a profound effect on twinning, but temperature has little or none. A wire stretched at high temperatures has fewer twin lamellae than one stretched at low temperatures, but this is due to the greater ease with which ordinary gliding occurs at high temperatures.

15. Work hardening.

As a single crystal is progressively deformed it becomes increasingly resistant to further deformation—a process known as 'work hardening', or 'cold working'. A single crystal cadmium wire, for example, after it has been plastically extended to twice its length will often support five times the load that originally caused it to deform, even though its area of cross-section has been reduced to a half of its original value. If, however, during the process of stretching the load is removed for a few minutes, extension may be produced subsequently with a smaller load than was required immediately before removing the load. This weakening of the crystal is a function of the time, for if the crystal is allowed to remain unloaded for about a day, its strength will approximate to its original small value.† The phenomenon is known as 'recovery' (German 'Kristallerholung'). These two opposing tendencies depend very much upon the temperature. It is impossible to work harden a substance near its melting point (which is the reason why lead will not harden at room temperatures), and equally impossible to observe recovery too far below its melting point.

The hardening observed on plastic deformation is due to the combination of two factors. As plastic deformation

† P. P. Ewald, Th. Pöschl and L. Prandtl, *The Physics of Solids and Fluids*, p. 127 (Blackie and Son, London, 1936).

proceeds the orientation of the glide elements relative to the direction of extension is continually changing, giving rise to an increase in the stress required to produce further deformation. But in addition to this the critical shearing stress itself also increases. If, for instance, a new pair of glide elements comes into play as a result of the reorientation, the critical stress required to bring about gliding on them is as great as or greater than the stress required at the end of the gliding on the previous pair. Thus it is seen that the latent glide elements are hardening at the same time as those actually effective during the deformation.

16. Relation of plastic deformation to structure.

The relation between plastic deformation and structure is obscure. No connection has been traced between twinning and structure, and only in the cubic face-centred metals and hexagonal metals is the glide plane the plane with the greatest density of atoms and the glide direction the direction of greatest line-density of atoms. In less symmetrical metals these relations do not hold. Calculations of the shearing stress under which ideal lattices would be expected to yield lead to a value about one-tenth that of the rigidity modulus. From the experimental evidence at present available it seems that for an ideal crystal the shearing stress required to produce plastic flow is about one-thousandth of that calculated on the basis of present theories of the mechanics of crystal lattices.

17. Fracture by cleavage.

When stress is suitably applied to crystals with a cleavage, particularly at low temperatures, fracture tends to take place on the cleavage planes exposing smooth and bright faces. When this occurs it is found that the stress per unit area normal to the cleavage plane which must be applied to produce fracture is a characteristic constant for the substance. This relation, which is fairly regularly satisfied, is known as Sohncke's law. If Z is the total stress per unit area along the rod, and χ is the inclination of the cleavage plane to the axis, then the stress per unit area of the cleavage plane in the direction of the rod is $Z \sin \chi$. The component of this normal to the cleavage plane is $Z \sin^2 \chi$.

If ρ_c is the critical normal stress

$$\rho_c = Z \sin^2 \chi,$$

and, according to Sohncke's law, ρ_c is a constant for a particular substance. The actual variation of Z with χ follows the law quite closely in the cases where it has been examined. The critical normal stress for zinc, bismuth, antimony and tellurium is of the order of 1 kg./mm.^{2} and is not greatly affected by previous cold working.

18. Relation between cleavage and crystal structure.

Haüy enunciated the earliest theory of cleavage based on structure, namely, that the ultimate particles of all crystals were bounded by cleavage faces. This theory is untenable because some cleavage blocks, e.g. fluorspar octahedra, cannot fill space. The next theory, based on Bravais' work on lattices, supposed that cleavage occurred in the crystal parallel to the plane with the maximum spacing. Though this is often true, it is by no means invariably so. Cleavage has been correlated with crystal structure† and a series of empirical rules enunciated. These may be briefly summarised as follows. All cubic ionic compounds with the general formula AX have cleavages parallel to {100}. Cleavages never break up radicals in ionic crystals, nor molecular complexes in homopolar crystals. Within this limitation the spacing of the planes parallel to the cleavages is a maximum. In ionic crystals which have no radicals cleavage occurs so as to expose planes of anions, where such planes exist; where they do not, large cations may also be exposed on the cleavage surface. Where several cleavages appear possible, they will all occur if their spacings are nearly the same, but if there is a

† N. Wooster, *Sci. Prog.* No. 103, p. 462 (1932).

considerable disparity only the largest will correspond to a cleavage.

Cubic crystals.

For those cubic crystals of the general chemical formula AX the cleavage is parallel to {100}. This is true both for the caesium chloride and sodium chloride type. The planes of ions which are farthest apart in the CsCl and NaCl types of structure are {110} and {100} respectively—a fact which shows clearly that there is no necessary connection between the planes of greatest spacing and the cleavage planes. In fluorspar, CaF_2, the cleavage plane is {111}, and planes parallel to

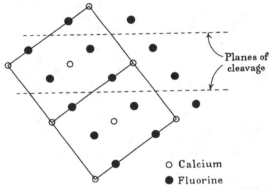

Fig. 25. Cleavage of fluorspar parallel to (111) planes.

this may be drawn through the structure so that their immediate neighbours are oxygen ions only, i.e. the cleavage exposes net planes of anions. This is shown in fig. 25, which is a projection of the structure with (111) perpendicular to the plane of the paper. In cuprite, Cu_2O, oxygen ions also occur immediately above and below the cleavage plane.

A good example of a homopolar crystal is diamond, for which the octahedral cleavage plane is the one with the greatest spacing. In the similar structure of zinc-blende, ZnS, the cleavage plane is {110}. It should be noted that a plane parallel to (111) in zinc-blende has only zinc atoms on one side, and sulphur atoms on the other, whereas (110) has an equal number of zinc and sulphur atoms on either side. Arsenious oxide, As_2O_3, is an example of a simple molecular structure. Its cleavage plane is $\{111\}$, and this is the only plane which can be drawn through the structure so as to pass through the crystal without cutting through the molecules of As_4O_6.

Isosthenic lattices.

In crystals of this type which contain radicals such as SO_4, SiO_4, the cleavage planes separate but do not cut through the radicals. Fig. 26 shows two projections of barytes, $BaSO_4$,

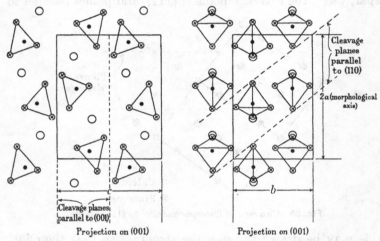

Fig. 26. Cleavages of barytes parallel to (001) and (110) planes.

and the dotted lines with the symbols attached indicate the cleavage planes. In potassium periodate, KIO_4, the cleavage plane (001) is bounded on either side by IO_4 radicals.

Layer lattices.

Without exception the cleavage planes of layer lattices are parallel to the planes of the layers in the crystal structure. Examples of this are afforded by graphite C (see fig. 27), zinc, cadmium, arsenic, antimony, bismuth, molybdenite MoS_2, cadmium chloride $CdCl_2$, cadmium iodide CdI_2, gypsum $CaSO_4.2H_2O$, mica $(H,K)AlSiO_4$, naphthalene $C_{10}H_8$ and many aromatic compounds containing flat molecules.

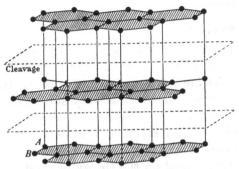

Fig. 27. Cleavage of graphite parallel to (0001) planes. (Reproduced from *Strukturbericht*, P. P. Ewald and C. Hermann (Akademische Verlagsgesellschaft, Leipzig, 1931), by kind permission of the authors and publisher.)

Chain lattices.

The cleavage planes are parallel to the chains. Examples of this are afforded by selenium, cinnabar HgS, and certain types of silicate structure. Thus the fibrous amphiboles and zeolites have chains of SiO_4 groups—simple in the amphiboles but more complex in the zeolites—running parallel to the cleavage planes. The cleavage plane of tremolite, $H_2Ca_2Mg_5(SiO_3)_8$, is parallel to $\{110\}$, and the 'plane' in the structure which is parallel to $\{110\}$ and does not cut through the SiO_4 groups is not quite flat but ridged. There would therefore appear to be a certain permissible departure from strict flatness in the structural planes corresponding to cleavage. In orthoclase the well-marked $\{001\}$ and $\{010\}$ cleavages are parallel to the folded chains† of AlO_4-SiO_4 four-membered rings.

Lattices with three-dimensional frameworks.

In a good example of this type of structure, namely quartz, there is no cleavage at all. Where cleavages occur in crystals which from their optical or other physical properties belong

† W. H. Taylor, *Z. Krist.* 85, 425 (1933), and *Proc. Roy. Soc.* A, **145**, 80 (1934).

to this group, e.g. the felspars and zeolites, there is, in those examples which have been examined, some characteristic of the structure which departs from that of a true three-dimensional framework.

19. Brittle fracture.

Brittle rupture and plastic flow are not to be sharply distinguished, since the time factor is of prime importance: some substances which are usually considered brittle may be plastically deformed when the force is very slowly applied, while others which appear plastic can show brittle fracture if the force is applied explosively. The breaking stress of crystals is found experimentally to be much less than that obtained from calculations based on the theory of crystal lattices. An estimate of the theoretical breaking stress of ionic crystals may be made by considering what force, Z, would separate two neighbouring planes perpendicular to a crystal rod of unit cross-section. If the assumption is made that the tension applied is proportional to the extension produced, right up to the breaking point, the work done in separating the two planes which form the new surfaces is $\frac{1}{2}Z\,\partial l$, where ∂l is the separation at rupture. Two new surfaces, each of unit area and surface tension S, have been produced, and the energy required to create these is $2S$. It is reasonable to assume that these quantities are approximately equal, so that

$$\tfrac{1}{2}Z\,\partial l \approx 2S.$$

S is known from other measurements and a limit may be set to ∂l as follows. It is reasonable to suppose that when the faces are separated by more than a few A.U. rupture occurs, because it is known that the attraction exerted by planes of atoms in crystals is inappreciable over a greater range than this. For rock-salt the value of Z calculated in this way is 200 kg./mm.^{2} Another line of reasoning shows that the breaking stress should be about one-tenth of Young's modulus, which would lead to about the same result for rock-salt. The observed breaking strength for dry rock-salt varies from one specimen to another, but is usually given as 0·4 kg./mm.^{2} This large discrepancy between the calculated and observed strength is typical of substances which break in a brittle manner.

20. Real and ideal crystals.

This discrepancy between theory and experiment has led to much discussion as to the nature of 'real' as opposed to 'ideal' crystals. Possibly much of the discrepancy is to be attributed to the existence of small superficial cracks. Two experiments which have a bearing on this explanation may be cited. A cleavage sheet of mica is cut into a strip of the same width as a pair of clamps. On being extended in these clamps the strip is found to have a low breaking strength. If the strip is now replaced by a wider piece, and the stress applied almost entirely at the central region by suitably designed clamps as shown in fig. 28, a much greater breaking stress is observed, about ten times the usual value.† It is supposed that this increase in strength is due to the impossibility of the surface cracks spreading inwards owing to the absence of stress at the edges. Specimens of rock-salt placed in water were found during solution to reach a breaking strength of about half the theoretical value.‡ This was alleged to be due to the solution and obliteration of the cracks on the surface. Some doubt has been cast on this explanation, however, as it has been suggested that the submerged rock-salt should not be re-

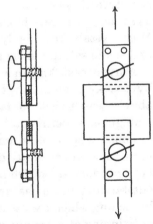

Fig. 28. Apparatus used to apply stress at the centre only of a sheet of mica.

garded as rock-salt simply, but rock-salt with absorbed water, this absorption producing a distortion of the lattice, which causes an increase of strength.

† E. Orowan, *International Conference on Physics.* Vol. ii, *The Solid State of Matter,* p. 89 (London Physical Society, 1935).

‡ A. F. Joffé, *The Physics of Crystals,* p. 62. (McGraw-Hill, New York and London, 1928.)

The existence of cracks in the crystal is supposed, according to some theories, to be an integral part of the make-up of real crystals. It is pointed out that from X-ray evidence there are two kinds of crystals—mosaic and perfect. Planes of the latter reflect X-rays over a few seconds of arc, the former over angular ranges up to a couple of degrees. The mosaic crystal was imagined to be built up of a number of small blocks of perfect crystal, of not more than some 500 A.U. side, arranged nearly parallel to each other. Theories have been put forward purporting to show that the intervening space is occupied by amorphous material, but in some cases at least the facts are not incompatible with a 'dry-built' solid formed of perfect but irregularly shaped blocks.

Many crystals, instead of growing as one solid body, grow as slightly deformed filled-in dendrites. This is known as 'lineage structure'.† The real crystal is not, therefore, to be regarded as so many discrete blocks, but rather as a kind of solid tree. Most crystals are unlikely to grow perfect because of impurities and strain set up during growth.

A hypothesis known as 'secondary structure', put forward by Zwicky in an attempt to explain the difficulties mentioned above, was based on faulty mathematics and has no unambiguous experimental justification.

Thus, in spite of an immense amount of research that has been done on the study of plastic deformation, no theory has been put forward which adequately explains all the facts. The outstanding problems are (1) the low elastic limit, (2) the hardening which results from deformation, (3) the appearance of twinning at a certain stage in the deformation.

A theory depending on the mutual repulsion of centres of stress explains many of the phenomena, but by no means all.‡ An account of most of the current theories is given in the report of a conference.§

† M. J. Buerger, *Z. Krist.* **89**, 195 (1934).
‡ G. I. Taylor, *Proc. Roy. Soc.* A, **145**, 362 (1934).
§ *International Conference on Physics.* Vol. II, *The Solid State of Matter* (London Physical Society, 1935).

CHAPTER III

CONDUCTION

1. Introduction.

The theory of thermal conduction in crystals was first studied about a century ago, and since that time the theoretical development has always been ahead of the experimental. The classical theory of conduction was completed fifty years ago, but up to the present there are no more than a dozen non-cubic crystals for which the absolute principal thermal conductivities are known. As well as the difficulty of obtaining large homogeneous crystals, there is the experimental difficulty of overcoming temperature drop where the heat enters and leaves the crystal. The methods of measurement that have been developed for isotropic conductors can be applied to crystal plates, provided they have been properly oriented with respect to the principal axes of the conductivity ellipsoid.

Investigations into electrical conductivity in crystals began rather later than the corresponding thermal investigations. Before the production of single crystals of metals very little experimental work was possible and even now the experimental data are nearly as meagre as for thermal conductivity.

There has recently been an enormous amount of study devoted to the properties of insulators. The development of electrical transmission systems carrying current at very high voltages has necessitated the systematic study of the conduction of this class of substances. It has great technical importance, but so far the welter of facts has not been sufficiently systematised to be included here. The same is true of the photoelectric and rectifying properties of certain crystals.

2. General theory.

Most of the general theory of conduction in crystals applies equally well to the conduction of heat and of electricity. Experiment has shown that in an isotropic conductor the flow, whether of heat or electricity, across unit area in unit time, is proportional to the gradient of temperature or potential in the direction of the flow.

For an isotropic substance the above relation between flow and potential may be expressed,

$$h = k\frac{d\theta}{dr},$$

where h is the amount of heat or electricity crossing unit area perpendicular to the direction of flow, in unit time, k is a constant known as the conductivity, $d\theta/dr$ is the temperature or potential gradient. h and $d\theta/dr$ are vector quantities, and if resolved into their components parallel to the three mutually perpendicular axes X_1, X_2, X_3, we obtain

$$h_1 = k\frac{d\theta}{dx_1}, \quad h_2 = k\frac{d\theta}{dx_2}, \quad h_3 = k\frac{d\theta}{dx_3}.$$

The correct extension of these relations to anisotropic substances is found to be

$$\left.\begin{aligned}
h_1 &= k_{11}\frac{d\theta}{dx_1} + k_{21}\frac{d\theta}{dx_2} + k_{31}\frac{d\theta}{dx_3} \\
h_2 &= k_{12}\frac{d\theta}{dx_1} + k_{22}\frac{d\theta}{dx_2} + k_{32}\frac{d\theta}{dx_3} \\
h_3 &= k_{13}\frac{d\theta}{dx_1} + k_{23}\frac{d\theta}{dx_2} + k_{33}\frac{d\theta}{dx_3}
\end{aligned}\right\} \qquad \ldots\ldots(1)$$

Equations (1) may be expressed in words as follows. Each component of the flow is a linear function of the components of the potential gradient. The nine constants are called conductivity constants. It is sometimes convenient to express the potential gradient in terms of the resistance coefficients and the flow. The corresponding equations are

$$\begin{aligned}
\frac{d\theta}{dx_1} &= r_{11}h_1 + r_{21}h_2 + r_{31}h_3 \\
\frac{d\theta}{dx_2} &= r_{12}h_1 + r_{22}h_2 + r_{32}h_3 \\
\frac{d\theta}{dx_3} &= r_{13}h_1 + r_{23}h_2 + r_{33}h_3
\end{aligned} \right\} \qquad \ldots\ldots(2)$$

From the form of equations (1) and (2) it will be seen that the k's and r's are second order tensors (cf. Chapter I). Equation (2) is the form in which the relation between current and potential is often expressed in electrical problems. In a later paragraph the relation between the k's and r's will be worked out. If we denote by k_{lm} all the nine k's, then $k_{lm} = k_{ml}$. No proof of this will be given here because the analysis is somewhat complicated, but the result of it is to show that if k_{lm} were not equal to k_{ml} then heat would flow out from a given point in spirals, not in straight lines. Experimental test both on the flow of heat and the flow of electricity has never shown any indication of non-linear flow and hence we may say that $k_{lm} = k_{ml}$. This condition reduces the number of k's and r's to six, namely k_{11}, k_{22}, k_{33}, k_{23}, k_{31}, k_{12}. The values of the k's depend on the orientation of the axes of reference X_1, X_2, X_3 with respect to the crystallographic axes, and in a later paragraph we shall show that if the axes X_i are properly oriented with respect to the crystallographic axes only three k's remain, namely, k_{11}, k_{22}, k_{33}, the others becoming zero. These three remaining k's are called the *principal* conductivity coefficients and are denoted k_1, k_2, k_3.

3. Isothermal and equipotential surfaces.

In the study of heat flow and electric current all places at the same temperature or same potential are supposed joined by continuous surfaces known as isothermal or equipotential surfaces respectively. If heat flows outwards from a point in an isotropic body or in a crystal belonging to the cubic system, the isothermal surfaces are spheres concentric with the source of heat. This must be so because if the body is isotropic the temperature distribution along any line passing through the source must be the same, no matter what its orientation relative to the coordinate axes. In anisotropic bodies the isothermal surfaces are ellipsoids. As we have seen in Chapter I, p. 13, the kind of ellipsoid and its orientation relative to the crystallographic axes is determined by the crystal symmetry.

4.* The two principal methods of measurement.

There are two important methods by which thermal or electrical conductivities may be measured. In one a flat plate of the crystal is clamped between two equipotential surfaces maintained at a constant difference of potential, while in the

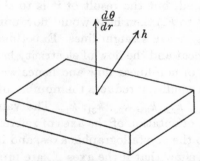

Fig. 29. Showing the relation between the directions of maximum temperature gradient and of the heat flow when opposite faces of a thin plate are maintained at different temperatures.

other the current flows along a rod of area of cross-section small compared with its length.

I. In the former method the equipotential surfaces are all necessarily parallel to the large faces of the plate. Just near the edges this is not true, but provided the thickness is small compared with the other dimensions, this will not introduce any appreciable error. The orientation of the plate therefore defines the direction of $d\theta/dr$. If we suppose the X_3' axis is normal to the plate, then

$$\frac{d\theta}{dx_1'} = \frac{d\theta}{dx_2'} = 0.$$

Hence from equation (1) we have

$$h_1' = k_{31}'\frac{d\theta}{dx_3'}, \quad h_2' = k_{32}'\frac{d\theta}{dx_3'}, \quad h_3' = k_{33}'\frac{d\theta}{dx_3'}.$$

The direction of heat flow in the plate does not therefore, in general, coincide with the normal to the plate, and if the material bounding it is isotropic the isothermal surfaces and directions of flow sufficiently far removed from the edges are as indicated in fig. 30.

It is clear that the total heat flowing normal to the section is h_3', since h_1' and h_2' do not carry any heat in that direction. Thus the quantities measured in this experiment are h_3' and

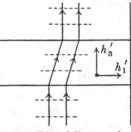

Fig. 30. Dotted lines are isothermal surfaces; full lines are lines of heat flow.

$d\theta/dx_3'$ and the coefficient k_{33}' is therefore obtained. The variation of k_{33}' with direction is given by the expression

$$k_{33}' = c_{13}^2 k_1 + c_{23}^2 k_2 + c_{33}^2 k_3, \qquad \ldots\ldots(3)$$

where k_1, k_2, k_3 are the principal conductivities and c_{13}, c_{23}, c_{33} are the direction-cosines of the direction of k_{33}' relative to the principal axes (see Chapter I). At least three differently orientated crystal plates are in general necessary for determining k_1, k_2 and k_3.

II. In the second method the current is passed through a rod, the cross-section of which is small compared with its length. The equipotential surfaces are not, in general, normal to the length of the conductor, and the relation of h and $d\theta/dr$ is shown in fig. 31. If we suppose the axis X_3' to be parallel to the length of the conductor, then $h_1' = h_2' = 0$ and $h = h_3'$. It is therefore more convenient to use equation (2), p. 65, from which we obtain

$$\frac{d\theta}{dx_1'} = r_{31}' h_3', \quad \frac{d\theta}{dx_2'} = r_{32}' h_3', \quad \frac{d\theta}{dx_3'} = r_{33}' h_3.$$

Usually the potential gradient along the conductor, $d\theta/dx_3'$, is measured and also the total flow h_3'; hence the coefficient derived is a resistance coefficient r_{33}', not a conductivity coefficient. The variation of r_{33}' with direction is given by

$$r_{33}' = c_{13}^2 r_1 + c_{23}^2 r_2 + c_{33}^2 r_3, \quad ...(4)$$

where r_1, r_2, r_3 are the principal resistivities and c_{13}, c_{23}, c_{33} are the direction-cosines of the direc-

Fig. 31. Showing the relation between the directions of heat flow and maximum temperature gradient when heat is caused to flow along a narrow rod.

tion of r_{33}' relative to the principal axes. At least three differently orientated rods are in general necessary to determine r_1, r_2, and r_3.

5.* The relation between conductivity and resistance coefficients.

In equation (1), p. 64, the components of flow are expressed in terms of the components of temperature or potential gradient. Conversely, in equation (2), p. 65, the components of temperature or potential gradient are expressed in terms of the components of flow. Writing in equation (2) the values of h_i in equation (1), we have

$$\frac{d\theta}{dx_1} = r_{11}\left(k_{11}\frac{d\theta}{dx_1} + k_{21}\frac{d\theta}{dx_2} + k_{31}\frac{d\theta}{dx_3}\right)$$
$$+ r_{21}\left(k_{12}\frac{d\theta}{dx_1} + k_{22}\frac{d\theta}{dx_2} + k_{32}\frac{d\theta}{dx_3}\right)$$
$$+ r_{31}\left(k_{13}\frac{d\theta}{dx_1} + k_{23}\frac{d\theta}{dx_2} + k_{33}\frac{d\theta}{dx_3}\right), \quad ......(5)$$

and similar expressions for $d\theta/dx_2$ and $d\theta/dx_3$.

Since this expression holds for any value of the potential gradient we may equate the coefficients of $d\theta/dx_1$, $d\theta/dx_2$, $d\theta/dx_3$; hence

$$\left. \begin{aligned} r_{11}k_{11}+r_{21}k_{12}+r_{31}k_{13} &= 1 \\ r_{11}k_{21}+r_{21}k_{22}+r_{31}k_{23} &= 0 \\ r_{11}k_{31}+r_{21}k_{32}+r_{31}k_{33} &= 0 \end{aligned} \right\} . \qquad \ldots\ldots(6)$$

If the three simultaneous equations (6) be solved in the usual way, we obtain

$$r_{11} = \frac{\begin{vmatrix} k_{22} & k_{23} \\ k_{32} & k_{33} \end{vmatrix}}{\begin{vmatrix} k_{11} & k_{12} & k_{13} \\ k_{21} & k_{22} & k_{23} \\ k_{31} & k_{32} & k_{33} \end{vmatrix}}, \quad \text{etc.,} \qquad \ldots\ldots(7)$$

where the vertical strokes indicate the determinants formed from the quantities within them. Proceeding in this way we may write down the values for all r_{ji}'s in terms of k_{lm}'s. The general expression is

$$r_{ji} = \frac{b_{ji}}{B}, \qquad \ldots\ldots(8)$$

where b_{ji} is the algebraic subdeterminant (minor) of the k_{lm}'s formed by omitting the jth row and the ith column, and B is the determinant of the k_{lm}'s.

Conversely, by substituting the expressions for $d\theta/dx$ from equation (2) in equation (1) the expression for k_{lm} in terms of r_{ji} is obtained. The expressions corresponding to (7) and (8) are

$$k_{11} = \frac{\begin{vmatrix} r_{22} & r_{23} \\ r_{32} & r_{33} \end{vmatrix}}{\begin{vmatrix} r_{11} & r_{12} & r_{13} \\ r_{21} & r_{22} & r_{23} \\ r_{31} & r_{32} & r_{33} \end{vmatrix}}, \qquad \ldots\ldots(9)$$

or, with due regard to the sign of a_{ji},

$$k_{ji} = \frac{a_{ji}}{A},$$

where a_{ji} is the subdeterminant obtained by omitting the jth row and the ith column from A the full determinant of the r's.

When the axes of reference coincide with the principal axes of the ellipsoid representing k'_{33} or r'_{33}, then

$$r_{12} = r_{21} = r_{23} = r_{32} = r_{31} = r_{13} = 0$$

(see Chapter I, p. 10), and expression (9) gives

$$k_{11} = \frac{r_{22} \cdot r_{33}}{r_{11} \cdot r_{22} \cdot r_{33}} = \frac{1}{r_{11}} \quad \text{and} \quad k_{22} = \frac{1}{r_{22}}, \quad k_{33} = \frac{1}{r_{33}}.$$

The ellipsoid corresponding to k'_{33} deduced in paragraph 4 may therefore be expressed in terms of the direction-cosines c_{13}, c_{23} and c_{33}, thus

$$k'_{33} = \frac{c_{13}^2}{r_1} + \frac{c_{23}^2}{r_2} + \frac{c_{33}^2}{r_3}, \qquad \ldots\ldots(10)$$

and conversely, the ellipsoid corresponding to r'_{33} may be expressed

$$r'_{33} = \frac{c_{13}^2}{k_1} + \frac{c_{23}^2}{k_2} + \frac{c_{33}^2}{k_3}.$$

If we put $r'_{33} = 1/k$, then the last expression may be put in the form often given in text-books, viz.

$$\frac{c_{13}^2}{k_1} + \frac{c_{23}^2}{k_2} + \frac{c_{33}^2}{k_3} = \frac{1}{k}. \qquad \ldots\ldots(11)$$

A partially true statement in connection with equation (11) is often made, which runs: "If an ellipsoid be constructed with principal axes proportional to the square roots of the principal conductivities of the crystal, then the length of any radius vector is proportional to the square root of the conductivity in the direction of that radius vector."

It is clear from equation (11) that the length of any radius vector of such an ellipsoid is proportional to $\sqrt{k}$, i.e. to $1/\sqrt{r'_{33}}$. Now this quantity has a meaning defined by equation (2), p. 65, and if the conduction occurs along a narrow rod, $1/r'_{33}$ is equal to the heat flow along the rod divided by the temperature gradient along it. k may therefore be considered as a conductivity, though it is not equal to k'_{33}, as may be seen from equation (1). Thus the statement given above in inverted commas would be correct if it ran "each radius vector of the conductivity ellipsoid is proportional to the reciprocal of the square root of resistivity in the direction of that radius vector". The difference between this statement and the previous one is serious if the ellipsoid has markedly unequal axes. As an illustration consider a conductivity ellipsoid for which $k_1 = 1$, $k_2 = 2$, $k_3 = 3$, and a direction such that $c_{13}^2 = c_{23}^2 = c_{33}^2 = \frac{1}{3}$; then, from equation (11),

$$\frac{1}{3}\left(\frac{1}{1}+\frac{1}{2}+\frac{1}{3}\right) = \frac{1}{k},$$

$$\frac{1}{3}\cdot\frac{11}{6} = \frac{1}{k},$$

$$k = \frac{18}{11},$$

and the conductivity so calculated is 1·64. If, however, we require the conductivity, i.e. k'_{33} and not $1/r'_{33}$, we must use equation (10), putting $1/r_1 = 1$, $1/r_2 = 2$, $1/r_3 = 3$, whence

$$\frac{1}{3}(1+2+3) = k'_{33},$$

$$k'_{33} = 2\cdot 0.$$

From this example it will be clear that it is necessary to specify whether conductivity or resistivity is meant when dealing with these problems.

6. Number of measurements necessary in the various crystal systems.

The number of conductivity coefficients required to give the value of the conductivity in a given arbitrary direction in the various crystal systems has been considered in paragraph 3, p. 6. If the measurements afford only k'_{33} (see paragraph 4), then at least as many independent values of k'_{33} as there are unknown conductivity coefficients must be obtained experimentally. Taking into account the equality of k_{lm} and

k_{ml}, there are six constants in the triclinic, four in the monoclinic, three in the rhombic, two in the trigonal, tetragonal, and hexagonal, and one in the cubic, systems.

In general, sections of any orientation with respect to the crystallographic axes can be cut from crystals. For all crystal systems except the monoclinic and triclinic the orientation of the conductivity ellipsoid relative to the crystallographic axes is fixed by the symmetry, and there is therefore no difficulty in obtaining the appropriate sections. It is not necessary to have the sections cut perpendicular to the principal axes of the conductivity ellipsoid, but the analysis is easier if they are, since each conductivity is determined separately.

In the monoclinic system one axis of the conductivity ellipsoid must coincide with the diad axis, or, if only a plane of symmetry is present, the axis is perpendicular to this. Thus one of the three principal sections may be cut from the crystal without difficulty. Three sections cut parallel to the diad axis, but otherwise having an arbitrary orientation, make it possible to determine the principal conductivity coefficients for directions perpendicular to the diad axis. The method of calculating the coefficients from the experimental data is the same as that given in Chapter II, p. 25, for the analogous problem in thermal expansion.

7.* The form of isothermal and equipotential surfaces and the general equation for the flow of heat or electricity in crystals.

Suppose heat or electricity to be generated at a constant rate at a point O inside a crystal. The crystal is supposed of such a size that when equilibrium has been established the difference in temperature or potential between its surface and its surroundings is negligibly small. Consider a very small rectangular block one corner of which has coordinates (x_1, x_2, x_3) relative to an origin at O. The dimensions of the block are dx_1, dx_2, dx_3. If the total heat flow per unit area along the line joining the point (x_1, x_2, x_3) to the origin be u, the heat flow in the x_1 direction normal to the two parallel faces bounded by dx_2, dx_3 will be u_1 and $u_1 + \dfrac{\partial u_1}{\partial x_1} dx_1$. Similarly, the total heat flowing through the other pairs of faces will be

$$u_2 dx_3 dx_1 \quad \text{and} \quad \left(u_2 + \frac{\partial u_2}{\partial x_2} dx_2\right) dx_3 dx_1,$$

$$u_3 dx_1 dx_2 \quad \text{and} \quad \left(u_3 + \frac{\partial u_3}{\partial x_3} dx_3\right) dx_1 dx_2.$$

Because temperatures are stationary there can be no accumulation of heat in the small volume and, therefore,

$$\frac{\partial u_1}{\partial x_1} + \frac{\partial u_2}{\partial x_2} + \frac{\partial u_3}{\partial x_3} = 0. \qquad \dots\dots(12)$$

If the crystal is isotropic and the conductivity k, then

$$u_1 = k \frac{\partial \theta}{\partial x_1}, \quad \text{etc.}$$

Substituting these values in equation (12),

$$k \left(\frac{\partial^2 \theta}{\partial x_1^2} + \frac{\partial^2 \theta}{\partial x_2^2} + \frac{\partial^2 \theta}{\partial x_3^2}\right) = 0. \qquad \dots\dots(13)$$

It is obvious that the isothermal or equipotential surfaces in a cubic crystal must be spheres, for which the equation is

$$x_1^2 + x_2^2 + x_3^2 = \text{constant} = f(\theta). \qquad \dots\dots(14)$$

The solution of this equation which corresponds to decrease of temperature with increasing distance from the origin, and also satisfies equation (13), is

$$x_1^2 + x_2^2 + x_3^2 = \frac{Q}{\theta^2}, \qquad \dots\dots(15)$$

where Q is a constant. This result may be verified by inserting the second partial differentials of θ with respect to x_1, x_2 and x_3 in equation (13). From equation (15) it follows that the temperature at any point is inversely proportional to its distance from the origin.

For an anisotropic crystal the relation between u_i and

$\partial\theta/\partial x_i$ varies with the direction. If the axes of the k'_{33} ellipsoid have been appropriately chosen we have seen that

$$u_1 = k_1 \frac{\partial\theta}{\partial x_1}, \quad u_2 = k_2 \frac{\partial\theta}{\partial x_2}, \quad u_3 = k_3 \frac{\partial\theta}{\partial x_3},$$

and the new form of equation (13) is therefore

$$k_1 \frac{\partial^2\theta}{\partial x_1^2} + k_2 \frac{\partial^2\theta}{\partial x_2^2} + k_3 \frac{\partial^2\theta}{\partial x_3^2} = 0. \qquad \text{......(16)}$$

This equation may be reduced to the same form as equation (13) by changing the variables. Thus if we put

$$x_1 = \sqrt{k_1} \cdot m_1, \quad x_2 = \sqrt{k_2} \cdot m_2, \quad x_3 = \sqrt{k_3} \cdot m_3,$$

then $\dfrac{\partial^2\theta}{\partial m_1^2} = k_1 \dfrac{\partial^2\theta}{\partial x_1^2}$, and similar expressions hold for $\dfrac{\partial^2\theta}{\partial m_2^2}$ and $\dfrac{\partial^2\theta}{\partial m_3^2}$. Hence equation (16) becomes

$$\frac{\partial^2\theta}{\partial m_1^2} + \frac{\partial^2\theta}{\partial m_2^2} + \frac{\partial^2\theta}{\partial m_3^2} = 0. \qquad \text{......(17)}$$

As with equation (15), the appropriate solution of this is

$$m_1^2 + m_2^2 + m_3^2 = \frac{R}{\theta^2},$$

or
$$\frac{x_1^2}{k_1} + \frac{x_2^2}{k_2} + \frac{x_3^2}{k_3} = \frac{R}{\theta^2}. \qquad \text{......(18)}$$

Thus for θ constant, i.e. over an isothermal surface, the point (x_1, x_2, x_3) always lies on a triaxial ellipsoid having axes proportional to the square roots of the principal conductivities.

Heat flow in an isotropic body.

Having considered the shape of the isothermal surface it is necessary to calculate the amount of heat flowing through it. In isotropic crystals the heat flow is in the direction of the maximum temperature gradient. Let the distance of a point from the heat source be r, then

$$r^2 = x_1^2 + x_2^2 + x_3^2 = \frac{Q}{\theta^2}, \qquad \text{......(19)}$$

or
$$r = \sqrt{Q} \cdot \theta^{-1} \quad \text{and} \quad \frac{dr}{d\theta} = -\sqrt{Q} \cdot \theta^{-2}. \qquad \text{......(20)}$$

The total heat given out by the source H must be equal to that passing through the sphere of radius r. Now from the ordinary law of conduction $H = -k \cdot A \cdot d\theta/dr$, where $A =$ area through which heat is flowing. Therefore, from equation (20),

$$H = -k \cdot 4\pi r^2 \frac{d\theta}{dr} = k \cdot 4\pi r^2 \frac{\theta^2}{\sqrt{Q}}$$

$$= k \cdot 4\pi \frac{Q}{\theta^2} \cdot \frac{\theta^2}{\sqrt{Q}} = 4\pi k \sqrt{Q}.$$

Equation (15) can therefore be rewritten thus:

$$\sqrt{x_1^2 + x_2^2 + x_3^2} = \frac{H}{4\pi k\theta}. \qquad \ldots\ldots(21)$$

From this equation it follows that the temperature at any point in the isotropic body above that which obtains at an infinite distance from the source is directly proportional to the rate of supplying heat at the source.

Heat flow in an anisotropic crystal.

Since it is established experimentally that heat flows out from a point source in straight lines, a portion of the crystal bounded by a cone of small angle, OQR, terminating at the source of heat, fig. 32, can be regarded as isolated from the rest of the crystal. Since the direction of the flow is defined by the axis of the cone, and the normal to the isothermal surface does not necessarily coincide with it, the case we are considering is that denoted II in paragraph 4, p. 67. The relevant relation determining the heat flow is thus:

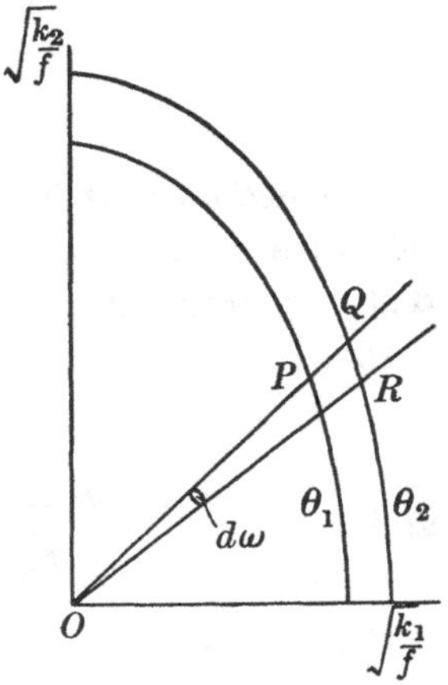

$$\frac{d\theta}{dx_3'} = r_{33}' h_3', \qquad \ldots\ldots(22)$$

Fig. 32

where $d\theta/dx_3'$ and h_3' are the temperature gradient and heat flow per unit area per unit time along the axis of the cone, and r'_{33} is the resistivity.

Equation (22) may be put into the form

$$h'_3 = k \frac{d\theta}{dx'_3},$$

where $k = 1/r'_{33}$. Since the ellipsoids corresponding to two neighbouring isothermal surfaces at temperatures θ_1 and θ_2 have principal axes in the same ratio, say x, the lengths OP and OQ are given by

$$\frac{c_{13}^2}{k_1} + \frac{c_{23}^2}{k_2} + \frac{c_{33}^2}{k_3} = \frac{1}{(OP)^2},$$

$$\frac{c_{13}^2}{x^2 k_1} + \frac{c_{23}^2}{x^2 k_2} + \frac{c_{33}^2}{x^2 k_3} = \frac{1}{(OQ)^2}.$$

Hence $$\frac{OQ}{OP} = x.$$

If the solid angle of the cone QOR is $d\omega$, the area of cross-section at Q is $OQ^2 d\omega$. If the total heat flowing through this cone is h,

$$h = OQ^2 d\omega \cdot h'_3$$

$$= OQ^2 k \frac{\theta_1 - \theta_2}{PQ} d\omega.$$

Now since no heat flows across the sides of the cone the temperature at any point will be inversely proportional to its distance from O, as shown above for an isotropic crystal. Hence

$$\frac{OQ}{OP} = \frac{\theta_1}{\theta_2}$$

and $$\frac{\theta_1 - \theta_2}{\theta_2} = \frac{OQ - OP}{OP} = \frac{PQ}{OP}.$$

Hence $$\frac{\theta_1 - \theta_2}{PQ} = \frac{\theta_2}{OP}$$

and $$h = OQ^2 k \frac{\theta_2}{OP} d\omega.$$

From equation (11) (p. 70) it follows that for all points on the θ_2 isothermal surface, k is proportional to OQ^2. If the factor of proportionality is f and the isothermal surfaces are brought very close together, then

$$h = f \cdot OQ^3 \cdot \theta_2 \cdot d\omega.$$

Now the volume of the cone $QOR = \frac{1}{3}OQ^3 \cdot d\omega = dv$. Hence

$$h = 3f\theta_2 \cdot dv.$$

Since f and θ_2 are the same for all points on the isothermal surface, the total heat, H, flowing out through it is given by

$$H = 3f\theta_2 \text{ (volume of ellipsoid)}.$$

The volume of the ellipsoid

$$= \tfrac{4}{3}\pi \sqrt{\frac{k_1 k_2 k_3}{f^3}}.$$

Hence $\qquad H = 3f\theta_2 \cdot \tfrac{4}{3}\pi \sqrt{\dfrac{k_1 k_2 k_3}{f^3}} = 4\pi\theta_2 \sqrt{\dfrac{k_1 k_2 k_3}{f}}$

or $\qquad\qquad \dfrac{H}{4\pi\sqrt{k_1 k_2 k_3} \cdot \theta_2} = \dfrac{1}{\sqrt{f}}.$

We may determine the value of f by considering the limiting case when $k_1 = k_2 = k_3 = k$, as in equation (21). From this it follows that

$$\frac{H}{4\pi\sqrt{k^3} \cdot \theta} = \sqrt{\frac{x_1^2}{k} + \frac{x_2^2}{k} + \frac{x_3^2}{k}} = \frac{1}{\sqrt{f}}. \qquad \ldots\ldots(23)$$

Thus f is a constant for any given isothermal surface and takes the form given in equation (23) for a cubic crystal. The corresponding equation for an anisotropic crystal is clearly

$$\sqrt{\frac{x_1^2}{k_1} + \frac{x_2^2}{k_2} + \frac{x_3^2}{k_3}} = \frac{1}{\sqrt{f}},$$

and hence we have finally

$$\frac{H}{4\pi\sqrt{k_1 k_2 k_3} \cdot \theta} = \sqrt{\frac{x_1^2}{k_1} + \frac{x_2^2}{k_2} + \frac{x_3^2}{k_3}}. \qquad \ldots\ldots(24)$$

This is the fundamental equation in the study of the flow of heat or electricity in anisotropic crystals.

8. Experimental methods of measuring thermal conductivity.

8.1. In metallic crystals.

The methods used for determining the principal conductivities of metallic single crystals do not call for special mention here. They are the standard methods used for polycrystalline materials and are to be found in the text-books of physics. Special researches at high or low temperatures with particular crystals have demanded slight modifications of the usual methods. In nearly all these experiments a constant source of heat was maintained at one end of a rod: measurements were made of the total flow of heat through the rod and the temperature difference over a measured length. If H is the total heat flow, θ_1, θ_2 the temperatures at points distant x from one another, A the area of cross-section of the rod, then the conductivity k is derived from the equation

$$H = kA\frac{\theta_1 - \theta_2}{x}.$$

The temperature gradient is easily measured, but various methods have been employed for the more difficult task of measuring the heat flow. The rate of supply of heat at one end of the rod may be readily found and maintained constant by employing electrical heating. This may be used directly to determine the heat flow provided there is no loss of heat from the sides of the rod. Usually it is difficult to prevent heat escaping and a check is made on the heat flow by absorbing and measuring the heat which has flowed down the rod. This heat flow is absorbed in a liquid and measured either by the rise in temperature of a known volume of liquid or, if the latter is volatile, by the volume of vapour produced at a given pressure.

8.2. In non-metallic crystals.

The low conductivity of these crystals makes the measurement rather more difficult than for metallic crystals. The classical divided bar method for poor conductors is available and is often used. The principle of the method is indicated in fig. 33. Two metal blocks of the same material are pressed against a crystal section having the same area and shape of cross-section as the metal blocks.

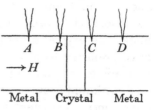

Heat is allowed to flow into one block and out of the other, and the conditions are controlled so that a steady state is reached. At least four thermocouples or other small thermometers, A, B, C and D, are used to measure the temperature gradients on either side of the crystal section. The couples B and C are placed as near to the crystal as possible, and from the curves drawn in fig. 33 one can

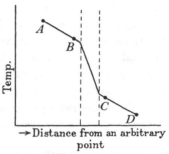

Fig. 33. Diagram showing (above) the arrangement of metal blocks on either side of the crystal plate and (below) the temperature drop along the composite bar.

extrapolate to the surface of the section and so obtain the temperature drop across it. The straight lines AB, CD should be parallel if the loss of heat from the surfaces of the blocks is small compared with the heat flowing through them. The slopes of the lines in fig. 33 are inversely proportional to the conductivities. This may be seen by putting H equal to the heat flowing through, k_m, k_c, the conductivities, and $(d\theta/dx)_m$, $(d\theta/dx)_c$, the temperature gradients in metal and crystal respectively. Then

$$H = k_m A \left(\frac{d\theta}{dx}\right)_m = k_c A \left(\frac{d\theta}{dx}\right)_c,$$

whence

$$\frac{k_m}{k_c} = \left(\frac{d\theta}{dx}\right)_c \bigg/ \left(\frac{d\theta}{dx}\right)_m.$$

The measurement is therefore one in which the conductivity of a crystal is compared with that of a metal of known conductivity. An experimental difficulty arises in establishing good thermal contact between the metal blocks and crystal. In general it is best to use a very thin liquid film between metal and crystal on both sides, though some experimenters have preferred to use no liquid. In any case allowance must be made for the inevitable temperature drop across the layer of air or liquid between the crystal and the metal. This may be done most conveniently by calibrating the apparatus with plates of known conductivity or using plates of the same material and orientation but of different thickness. It is worth noting that this method actually measures a coefficient of conductivity, whereas the rod method described for metal crystals determined a coefficient of resistance. Both of these methods are equally applicable to crystalline and amorphous materials; the next paragraph describes a method applicable only to anisotropic crystals.

9. Twin-plate method of measuring the ratio of two principal thermal conductivities.

When heat flows through a crystal in a direction not parallel to one of the principal conductivity axes the vectors representing the heat flow and the temperature gradient do not coincide. It is possible to derive the ratio of two of the principal conductivities from the angle between these two vectors.†

In fig. 34 OA and OB represent the directions of two of the principal conductivity axes

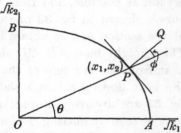

Fig. 34. The relation in a principal section of an isothermal surface between the direction of maximum temperature gradient PQ and the direction of heat flow OP.

† The following analysis also applies to the determination of the angle between the wave-normal and the ray in a uniaxial crystal.

and the axes of reference OX_1, OX_2: the lengths OA, OB are proportional to $\sqrt{k_1}$, $\sqrt{k_2}$, where k_1, k_2 are the conductivities in these directions. Heat flows outwards in all directions from O and an ellipse having OA and OB as its principal axes is a principal section of an isothermal surface (see paragraph 3). The maximum temperature gradient at a point P on this isothermal surface is necessarily directed along the normal PQ. If the coordinates of the point P are x_1, x_2 the equation to the ellipse is

$$\frac{x_1^2}{k_1} + \frac{x_2^2}{k_2} = \text{constant.}$$

The slope of the normal at P is therefore $\dfrac{x_2 k_1}{x_1 k_2}$.† If θ is the angle POA and ϕ the angle between the direction of the temperature gradient and OP, then

$$\tan\theta = \frac{x_2}{x_1} \quad \text{and} \quad \tan(\theta+\phi) = \frac{x_2 k_1}{x_1 k_2} = \frac{k_1}{k_2}\tan\theta.$$

When k_1 and k_2 are nearly equal, ϕ is small and we may write

$$d(\tan\theta) = \sec^2\theta \,.\, d\theta,$$

$$\tan(\theta+\phi) - \tan\theta = \sec^2\theta \,.\, \phi,$$

$$\tan\theta\left(\frac{k_1}{k_2} - 1\right) = \sec^2\theta \,.\, \phi$$

or

$$\phi = \tfrac{1}{2}\sin 2\theta \left(\frac{k_1}{k_2} - 1\right).$$

Thus if ϕ and θ can be measured k_1/k_2 may be calculated. Further, if k_1 and k_2 are nearly equal we see that ϕ is a maximum when $\theta = 45°$, i.e. when the direction of heat flow bisects the angle between the directions of the principal conductivities.

In order to measure ϕ easily a so-called twin-plate is constructed. A rectangular plate of the crystal containing two principal axes is prepared. This is represented in fig. 35 by

† S. L. Loney, *Coordinate Geometry*, p. 240 (Macmillan, London, 1920).

$ABCD$; k_1 and k_2 give the directions of the principal conductivities. The plate is then cut along the line, indicated by the dotted line in fig. 35, parallel to the side AB. The lower half is then turned over so that the edge CD occupies the previous position of the dotted line. The principal axes in the two halves are related in the same way as they would be by a twin-plane parallel to AB. The two halves of the original

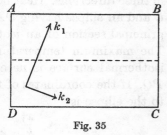

Fig. 35

plate are cemented together along the line DC, fig. 36. The end EF, fig. 37, of this 'twin-plate' is ground flat and cemented

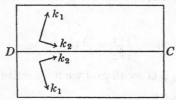

Fig. 36. The relation between the axes k_1, k_2 of the isothermal surface in the two halves of the twin-plate.

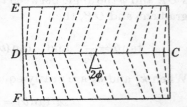

Fig. 37. Diagram showing the mutual inclination of the isothermal surfaces in the two halves of the twin-plate.

to a copper block which can be conveniently heated. The position of one isothermal surface may be found in the following way. A mixture of wax and elaidic acid in certain proportions is melted and spread in a thin layer over the twin-plate. This mixture has the property that on rapid cooling the grain size is fine and the appearance light, while on slow cooling the grains are coarse and the appearance dark. The melted layer is cooled rapidly so that at the beginning of the experiment the twin-plate appears light. Heat is introduced into the plate through the copper bar so that the wax melts slowly. Near the side EF of the plate the isothermal surfaces are almost parallel to the edge, but at a short distance from it they become inclined, as shown by the dotted lines in fig. 37.

The angle between the isothermal surfaces in the two halves along the centre line CD remains constant except near the ends C, D, and is measured near the centre of the plate because in this region disturbances due to loss of heat from the surface of the plate are a minimum. The copper plate is heated sufficiently to melt the wax to the desired extent. The whole plate is allowed to cool slowly so that the melted portion appears dark. The angle between the isothermal surfaces can easily be measured under these conditions and the value of 2ϕ obtained. The orientation of the central line of the twin-plate with respect to the principal axes of conductivity gives θ.

10. Measurement of the principal electrical conductivities in metallic crystals.

The methods of measuring the electrical resistance of metallic wires are too well known to require description here. The experimental difficulties only arise in securing good crystals having the required orientation. An account of some of the better known methods of growing single crystals of metals is

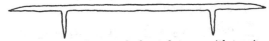

Fig. 38. Form of a single crystal of metal grown with two 'potential' leads as part of the crystal.

given in Appendix IV. It is easy to grow a single crystal wire with two 'potential' leads on it as shown in fig 38. This is generally done when the conductivity is sensitive to mechanical strain, because the clamping of potential leads on to the wire may otherwise lead to inaccurate results. The validity of the relations given in paragraph 4 has been established for a large number of metals, particularly for the hexagonal close-packed group, zinc, cadmium, magnesium, and for bismuth.

11. The relation between thermal conductivity and crystal structure.

The relation between thermal conductivity and crystal structure cannot as yet be properly studied because the data available are too fragmentary. The absolute values of the thermal conductivities are known only for a few of those crystals which are readily obtained in large pieces. The *ratios* of the principal conductivities are known for some fifty substances, but the measurements are not accurate and the substances seem to have been chosen at random. A correlation does appear to exist between thermal conductivity and crystal structure in layer and chain lattices. Table XV gives the data available for these types (Groups 2 and 3).

It will be seen that the layer lattices all show greater conductivity in the layers than perpendicular to them, and that chain lattices have the greatest conductivity along the chains. An attempt has been made† to associate the anisotropy of the thermal conductivity with the geometry of the crystal structure for those types of crystal where the bonds are approximately equivalent throughout. Only the bonds between atoms in contact were considered, and their relative lengths ignored. Each bond was defined by the angles it made with the principal axes. Thus for the ath bond these angles were θ_1^a, θ_2^a, θ_3^a. The quantities

$$a_1 = \sum_{1-n} \cos^2\theta_1^a, \quad a_2 = \sum_{1-n} \cos^2\theta_2^a, \quad a_3 = \sum_{1-n} \cos^2\theta_3^a,$$

when the summation was taken over all the bonds in the cell were found to have approximately the same ratios to one another as the principal conductivities k_1, k_2, k_3. This theory was tested on all the available data, but these are so few that it cannot be regarded as established (cf. Table XV, Group 1).

12. Thermoelectricity.

12.1. Introduction.

When two dissimilar metal wires are joined together to form a closed circuit an electric current flows round the circuit whenever the temperatures of the junctions of the two metals are different. Such a current obtained by heating a junction

---

† W. A. Wooster, *Z. Krist.* **95**, 138 (1936).

TABLE XV. *Anisotropy of thermal conductivity in relation to crystal structure*

Formula	Name	Structure type†	$k_\perp/k_\parallel$‡ Observed	Calculated
Group 1				
TiO_2	Rutile	C4	0.6_2§	0.8_3
Al_2O_3	Corundum	D51	0.8_5	0.7_4
$CaSO_4$	Anhydrite	H1	$\{k_1/k_3 = 0.9_4$ $\{k_2/k_3 = 0.8_9$	$0.9_7\}$ $1.00\}$
$ZrSiO_4$	Zircon	H3	0.8_1	0.8_4
KH_2PO_4	—	H22	1.4_5	1.1_0
$CO(NH_2)_2$	Urea	O21	0.7_9	0.5_2

Group 2 (Layer lattices)	Name	Structure type†	$k_\perp/k_\parallel$	k_1/k_3	k_2/k_3	Plane of layers
Zn	Zinc	A3	1.0_6∥	—	—	(0001)
Sb	Antimony	A7	2.5_3	—	—	(0001)
Bi	Bismuth	A7	1.3_9,¶ 1.7_3††	—	—	(0001)
C	Graphite	A9	4.0	—	—	(0001)
$CaSO_4 2H_2O$	Gypsum	—	—	0.6_4	0.4_2	(010)
$(H, K)AlSiO_4$	Mica	—	—	5.8	6.3	(001)

Group 3 (Chain lattices)	Name	Structure type†	$k_\perp/k_\parallel$	k_1/k_3	k_2/k_3	Direction of chains
Te	Tellurium	A8	0.6_6	—	—	[0001]
HgS	Cinnabar	B9	0.7_2	—	—	[0001]
TiO_2	Anatase	C5	1.8_0	—	—	[100]
SiO_2	Quartz	C8	0.5_8	—	—	[111] and [1Ī0]
Hg_2Cl_2	Calomel	D31	0.5_9	—	—	[001]
KH_2PO_4	—	H22	1.4_5	—	—	[100]
Sb_2S_3	Stibnite	—	—	0.4_7	0.2_9	[001]
$H_2Ca_2Mg_5(SiO_3)_8$	Tremolite	—	—	0.3_6	0.5_7	[001]

† See P. P. Ewald and C. Hermann, *Strukturbericht*.

‡ $k_\perp/k_\parallel$ gives the ratio of the conductivities perpendicular and parallel to the optic axis.

k_2 is conductivity parallel to the [010] axis in rhombic and monoclinic crystals.

k_1 and k_3 are the principal conductivities parallel to [100] and [001] axes in rhombic crystals and lie in (010) plane in monoclinic crystals.

§ Except where otherwise stated values of ratios of thermal conductivities are taken from the *International Critical Tables*, vol. v, p. 230.

∥ C. A. Cinnamon, *Phys. Rev.* **46**, 215 (1934).

¶ G. W. C. Kaye and J. K. Roberts, *Proc. Roy. Soc.* A, **104**, 98 (1923).

†† G. W. C. Kaye and W. F. Higgins, *Phil. Mag.* 8, 1056 (1929).

of, say, copper and iron wires to $100.^{\circ}$ C. whilst the other is at room temperature may easily be recorded on a table galvanometer. Crystals which are good electrical conductors behave in the same way as polycrystalline rods and wires. In addition to obtaining the thermoelectric effect with pairs of different crystals we may also obtain it by placing two crystals of the same anisotropic substance in contact, provided they are not similarly oriented. For example, if two cubes of haematite each cut with one face coinciding with the basal plane (0001) are placed in contact, the basal plane of the one, and the prism face of the other being together, then a difference of potential develops between the faces opposite to those in contact when the junction is heated. On joining the outer faces with a wire a current flows at the hot junction from the cube with its trigonal axis parallel to the common face, to the other crystal. The maximum current is obtained when the inclination of the axes of the two crystals is 90°, and decreases to zero when they are parallel. Since the thermoelectric power relates two vectors, namely, the temperature gradient and the potential gradient, its variation with direction is given by the usual expression for a second order tensor,

$$g'_{33} = c^2_{13}g_{11} + c^2_{23}g_{22} + c^2_{33}g_{33},$$

where g'_{33} is the thermoelectric power in an arbitrary direction defined by direction-cosines; c_{jk} and g_{ii} are the principal thermoelectric powers. As for other second order tensors, the thermoelectricity of triclinic, monoclinic and rhombic crystals may be represented by a triaxial ellipsoid; for hexagonal, trigonal and tetragonal crystals it may be represented by an ellipsoid of revolution, and the corresponding surface for cubic crystals is a sphere.

Thermoelectric properties are very sensitive to various physical and chemical disturbances of the lattice. Impurities, even in very small amounts, mechanical strain, magnetic fields, and temperature changes, all have profound effects on the thermoelectric e.m.f. A thorough systematic study of all known conducting crystals is, therefore, wellnigh impossible, because many of them are sulphides, which only occur naturally and are usually far from being chemically homogeneous. The following experiment indicates another of the difficulties in the experimental study of this subject. A crystal, held in a suitable clamp, is connected through an electrometer with a heated wire. If this wire be brought into contact with various adjacent points of a flat face of the crystal the potentials measured are very different. In some cases the current will change its direction, when the wire is moved a small distance. Under such circumstances, which fortunately are not general,

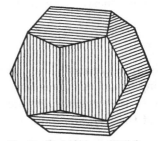

Fig. 39. Striations on {210} faces of iron pyrites which is thermo-electrically positive.

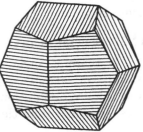

Fig. 40. Corresponding striations on thermoelectrically negative crystals.

measurements which are obtained with crystal plates are not reproducible. In certain sulphides there appears to be a relation between the striations to be seen on the crystal and the thermoelectric sign (which is defined with respect to the material of the electrodes, being called positive when at the cold junction the current flows from the crystal to the electrode). Thus many crystals of iron pyrites, FeS_2, are striated on the {210} faces parallel to the [001] direction, while others are striated in a direction which is perpendicular to this on the {210} planes, i.e. in the [$\bar{1}$20] direction. The thermoelectric sign of the former set is positive (like antimony), and the latter is negative (like bismuth) (figs. 39, 40). A commonly repeated incorrect assertion is to be found in the literature, that the striations in the first form are due to the appearance of faces of the form {210}, and in the second to faces of the form {120}. This has been shown to be false by X-ray analysis, which was able to demonstrate in some twenty crystals of both electric sign and both types of striations that the only form occurring was {210}. Why there should be this remarkable correlation between thermoelectric sign and the striations is, up to the present, completely obscure.

In comparing the various crystals it is generally true to say that the worse the conductivity the greater the thermoelectric power.† Thus molybdenite, MoS_2, and graphite, C, have very high thermoelectric powers compared with haematite, Fe_2O_3, but haematite has a very much higher conductivity than the other two.

12.2. General theory.

The thermoelectric power of a couple consisting of an isotropic metal and a crystal rod having its axis and end faces in a definite orientation with respect to the crystallographic axes is dependent on the orientation of the rod, but indepen-

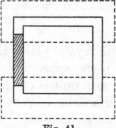

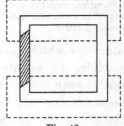

Fig. 41   Fig. 42

dent of the orientation of the end faces. In figs. 41 and 42 the unshaded part of the rectangular ring represents an isotropic conductor and the shaded part the crystal rod, the length of which bears the same relation to the crystallographic axes in both cases. Provided the junctions are completely enclosed in a constant temperature bath, indicated by the dotted lines,

† The thermoelectric power of a given couple is the e.m.f. developed when the temperature difference between the hot and the cold junctions is equal to 1° C.

there can be no difference in the e.m.f. produced. If there were a difference in thermo-e.m.f. in these two circuits a couple with one surface separating crystal and metal as in fig. 41 and one as in fig. 42 would give an e.m.f. even when at the same temperature. Useful work could therefore be abstracted from an isothermal enclosure. This would contradict the second law of thermodynamics, hence the original statement must be valid.

12.3. The heating effect which occurs when an electric current changes its direction in a crystal.

At the junctions of a thermocouple composed of two dissimilar isotropic metals, heat is produced or absorbed when current is sent round the couple from a battery or other source of electric power. This heat evolution and absorption is called the Peltier effect. The amount of heat generated is equal to the product of the current and a constant which depends on the nature of the two metals. Peltier heat is also evolved or absorbed when current is sent through two pieces of the same anisotropic crystal differently orientated with respect to the crystallographic axes. Using the two cubes of haematite mentioned on p. 86, the maximum Peltier effect is observed when the trigonal axes in the two blocks are perpendicular to each other.

The following analysis, due to Bridgman,† shows that a consequence of this result is that if a crystal be so cut or grown that the electric current has to change its direction in flowing round the crystal there is an evolution or an absorption of heat at all those places where the direction of the current changes. A single crystal cut in the ring-form shown in fig. 43 is divided into four segments and put together as shown in fig. 44. The direction of the principal axis in two segments is now perpendicular to its previous direction. The ring is placed in a constant temperature bath, and electric current caused to

† P. W. Bridgman, *The thermodynamics of electrical phenomena in metals*, p. 109 (Macmillan, New York, 1934).

flow, say, in a clockwise direction. At every junction the current flows from a crystal in which the principal axis is perpendicular to the current flow to one in which it is parallel to the current flow. If the Peltier effect for such a junction is an absorption of heat, then there will be such absorption at all the junctions. Now since the whole ring is in a constant temperature enclosure there can be no thermo-e.m.f. acting round the circuit. If we assume that apart from the Joule heating effect there is no heat evolution in between the

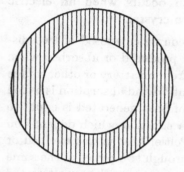 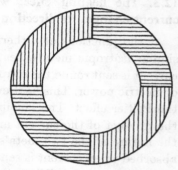

Fig. 43. Diagram illustrating a single crystal grown or cut into the form of a ring.

Fig. 44. Diagram illustrating a composite ring made from a single crystal ring by changing the orientation of two quadrants.

junctions heat is continuously abstracted from the constant temperature bath. If the battery were supplying work to abstract this heat it would be necessary at ordinary temperatures, from the laws of thermodynamics, for heat to be evolved somewhere else in the system: since the heat evolution in the battery is independent of the circuit attached to it this heat cannot appear in the battery—hence the original assumption is invalid. The only place where there could be a compensating heat evolution is in the portions of the sector between the junctions. In these portions of the circuit only the *direction* of the current is changing with respect to the principal axis of the crystal, and hence we establish this remarkable property of the flow of electricity through crystals.

12.4. Kelvin's axiom and the calculation of the general expression for the Peltier heat.

Suppose the crystal bar has its principal axis inclined to its length at an angle θ, as shown in fig. 45, and that it forms a junction with an isotropic metal. The heat developed (or absorbed) at the junction varies with θ, and is denoted P_θ. For a reason which will appear in the next paragraph we can proceed with the calculation of P_θ most easily by considering separately the effects of the

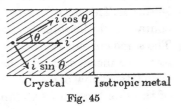

Crystal   Isotropic metal

Fig. 45

components of the current, i, parallel and perpendicular to the principal axis. The area of the junction projected on to a plane perpendicular to the component ($i\cos\theta$) is a fraction equal to $\cos\theta$ of the actual area of the junction. Hence the Peltier heat developed by the component of the current parallel to the principal axis is

$$i\cos\theta\,\cos\theta\,.\,P_{\parallel},$$

where $P_{\parallel}$ is the value of P_θ when $\theta = 0°$. Similarly, the Peltier heat developed by the other component is

$$i\sin\theta\,\sin\theta\,.\,P_{\perp},$$

where $P_{\perp}$ is the value of P_θ when $\theta = 90°$. Thus if we simply add the heating effects of the separate components

$$iP_\theta = i\cos^2\theta P_{\parallel} + i\sin^2\theta P_{\perp}$$

or $$P_\theta = (P_{\parallel} - P_{\perp})\cos^2\theta + P_{\perp}. \qquad \ldots\ldots(25)$$

This addition of the heating effects of the components perpendicular and parallel to the principal axis was introduced by Kelvin as an axiom. It has not been entirely justified on theoretical grounds though it is probably correct; its final justification depends on the verification by experiment of the relations derived by means of it. This particular relation for the variation of P_θ with θ has been roughly checked for bismuth and tin.

12.5. Transverse heating effect.

A crystal rod in which a principal thermoelectric axis is inclined to its length is maintained throughout at the same temperature. If an electric current is sent along the rod, there is an evolution of heat on one side and an equal absorption on the other. This may be seen by supposing the crystal rod to be in contact with the same isotropic metal on two sides (see fig. 46). The same current density is supposed to flow in the bounding isotropic metal and all conditions are arranged so that there is no lateral flow of electric current. The projection of unit area of the side face on to a plane perpendicular to $i \cos \theta$ is equal

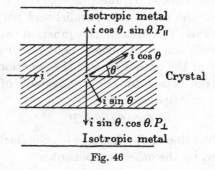

Fig. 46

to $\sin \theta$, and hence the Peltier heating due to this current flowing through unit area of the side face would be

$$i \cos \theta \sin \theta . P_{\parallel}.$$

Similarly, the Peltier cooling due to the other component flowing in the opposite direction through the side face is

$$i \sin \theta \cos \theta . P_{\perp},$$

and the difference gives the resultant transverse heating effect

$$T_{\theta} = \tfrac{1}{2} \sin 2\theta (P_{\parallel} - P_{\perp}).$$

T_{θ} is therefore zero for $\theta = 0°$ or $90°$ and a maximum when $\theta = 45°$. Thus for the particular components of i that were used in the previous section, there was no transverse heating effect, and it was for this reason that they were chosen.

If a current passes through a crystal rod which is not bounded on both sides by isotropic metal there will be, if θ is not equal to $0°$ or $90°$, a generation of heat on one side and an absorption on the other. This heat is dissipated by the cooling of the air, but even so the difference of temperature on the two sides of a bismuth rod 6 mm. in diameter and carrying a current of 1 ampere may amount to $0·5°$ C. It is therefore easy to demonstrate this effect and to check quantitatively the relation between the transverse heating and θ.

It follows as a thermodynamical consequence that there is an inverse effect. If the two sides of a crystal rod which are inclined to the principal axis at an angle other than $0°$ or $90°$ are maintained at different temperatures an electric potential difference is developed along the length of the rod.

12.6. Relation between the number of Peltier constants and the crystal symmetry.

From the form of equation (25) it is clear that a representation surface giving the variation of P_θ with θ would be of the second order. In fact we may write

$$\frac{1}{(1/\sqrt{P_\theta})^2} = \frac{\cos^2\theta}{(1/\sqrt{P_\parallel})^2} + \frac{\sin^2\theta}{(1/\sqrt{P_\perp})^2},$$

which is the equation to an ellipse.

The more general expression clearly derives from the experimental fact that between the Peltier heats developed on the three faces of a unit cube of the crystal, Q_l, and the three components of the electric current flowing through it, there is a linear relation of the form we have already met, namely,

$$Q_l = P_{kl} i_k.$$

The number of P_{kl}'s will vary with the crystal system in the way discussed on p. 7, and the representation surface having $1/\sqrt{P}$ as its radius vector will be a triaxial ellipsoid in the general case, an ellipsoid of revolution in uniaxial crystals, and a sphere in cubic crystals.

CHAPTER IV

INDUCTION

MAGNETIC INDUCTION

1. Introduction.

The power of attracting iron and the property of setting in a particular direction when freely suspended were characteristics of the loadstone which were described by Aristotle. But though ferromagnetism had been observed for centuries, diamagnetism and paramagnetism were discovered only in the middle of last century. A German botanist, Plücker, studied the effect of a magnetic field on pieces of wood and bark, and finding that a torque was exerted on them, he extended his investigations to crystals. Some, he found, had a tendency to become orientated with their length either parallel or perpendicular to the field. The theory to account for this difference was worked out shortly afterwards by W. Thomson (later Lord Kelvin).

If a rectangular parallelepiped of an isotropic substance is placed in a homogeneous magnetic field, magnetic poles are induced on the faces of the block perpendicular to the field. If the field strength is H, and the induced magnetic pole strength per unit area for this field is σ, then

$$\sigma = \chi H, \qquad \qquad \ldots\ldots(1)$$

where χ is a constant called the susceptibility. If the area of the end face is A, the total pole strength is σA. If the length of the parallelepiped is l, the total moment is $A\sigma l$. Thus σ is equal both to the magnetic moment per unit volume and to the pole strength per unit area.

Different authors use different units to express diamagnetic susceptibilities: χ may refer to unit volume, or unit mass, or may be expressed in terms of the gram-molecular weight (if χ_1 is the susceptibility for unit mass, and χ_m is the susceptibility per gram-molecular weight, then $\chi_m/\chi_1 = M$, the gram-molecular weight).

For theoretical discussion the unit used is generally immaterial, but in practical work it is essential to specify which meaning is intended. When χ is expressed in terms of the gram-molecular weight the value of the susceptibility for diamagnetic substances ranges from 0 to -300×10^{-6}, and for paramagnetic substances from 0 to $17,000 \times 10^{-6}$. For ferromagnetic substances the susceptibility is usually given per unit mass, and the values range from a few units to some 300.

If a ferromagnetic substance is placed in a uniform magnetic field the reaction between the two is so great that the field ceases to be uniform. With diamagnetic and paramagnetic substances, however, the degree of distortion of the field is so slight that the field remains almost uniform. Consequently, if an isotropic diamagnetic or paramagnetic substance is placed in a uniform field every part is equally affected, with the result that no turning force is exerted on the specimen. On the other hand, if the field is divergent, and the specimen rod-shaped, the field will in general exert a turning moment on the substance. If the rod is suitably suspended it will set with its length perpendicular to the field if it is diamagnetic, and parallel to it if it is paramagnetic. The reason for this may be illustrated by the following simple case.

Let two similar small elements in the rod be arranged symmetrically with respect to the suspension P (fig. 47). (The suspending fibre is supposed perpendicular to the plane of the paper.) Let the magnetic field radiate from the pole O. If the rod is diamagnetic A and B will experience a repulsion from the pole; the repulsion will be greater for A than for B because A is nearer. Hence the rod will set parallel to the dotted line (1). If the rod is paramagnetic, however, the pole O will exert an attraction on the rod, and as A is nearer than

B the end A will be the more strongly attracted; hence the rod will take up a position parallel to the dotted line (2).

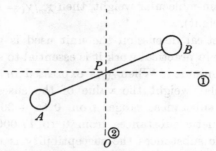

Fig. 47. Diagram indicating the origin of the twist experienced by paramagnetic and diamagnetic bodies in a non-uniform magnetic field.

With magnetically anisotropic crystals other effects are observed. If a sphere cut from such a crystal, suspended with the maximum paramagnetic susceptibility or minimum diamagnetic susceptibility perpendicular to the suspension, is placed in a magnetic field it sets with the direction of maximum algebraic susceptibility parallel to the field. When the anisotropic crystal has any other shape than spherical, and is placed in a non-uniform field, the anisotropy and the effect due to shape of the crystal interact. Much of the early work with diamagnetic substances was confused because the necessity of separating these two effects was not appreciated. Some substances behaved sometimes in one way, sometimes in another, depending in fact on the relative dimensions of the specimens tested.

The mechanical effects shown by a crystal in a magnetic field depend on the medium in which it is immersed. Consider (see fig. 48) a paramagnetic crystal of any form which is surrounded by a vacuum and a paramagnetic liquid in which there is a hole filled with vacuum of exactly the form of the

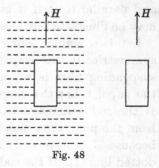

Fig. 48

crystal. If the same magnetic field acts on both the crystal
and the liquid the poles induced on the end faces of the
hole will be opposite to those induced on the crystal. When
the crystal occupies the hole the resultant pole strength will be
equal to the difference between the pole strengths induced in
the liquid and on the crystal. These pole strengths are proportional to the susceptibilities of crystal and liquid respectively, and the net effect of the magnetic field on the crystal
placed in a liquid is therefore proportional to the difference
between their susceptibilities.

2.* General theory.

*Relations between the induced magnetic moment and the
magnetising field.*

The induced magnetic moment in diamagnetic and paramagnetic substances is found to depend only on the first power
of the components of the magnetising field. Thus, as was
indicated in Chapter I, the three components of the magnetic
moment M per unit volume may be expressed:

$$M_1 = \chi_{11}H_1 + \chi_{21}H_2 + \chi_{31}H_3,$$
$$M_2 = \chi_{12}H_1 + \chi_{22}H_2 + \chi_{32}H_3,$$
$$M_3 = \chi_{13}H_1 + \chi_{23}H_2 + \chi_{33}H_3,$$

or, in tensor notation, $M_i = \chi_{ki}H_k$.

The quantities χ_{ki} are coefficients which depend on the orientation of the coordinate axes relative to the crystallographic
axes. It is shown in Appendix II that $\chi_{ik} = \chi_{ki}$, and therefore
the relations between M and H are formally the same as
those between the heat flow and the temperature gradient
(Chapter III). Thus, the magnetic moment in any direction
may be obtained by transforming to a system of axes X_1', X_2',
X_3' and considering the new X_3' axis to coincide with this
direction. Then, as shown on p. 13, we have

$$\chi_{33}' = c_{13}c_{13}\chi_{11} + 2c_{13}c_{23}\chi_{12} + 2c_{13}c_{33}\chi_{13}$$
$$+ c_{23}c_{23}\chi_{22} + 2c_{23}c_{33}\chi_{23}$$
$$+ c_{33}c_{33}\chi_{33}. \quad \ldots\ldots(2)$$

This expression is the general equation of a triaxial ellipsoid, and if the coordinate axes be chosen appropriately we may obtain

$$\chi'_{33} = c^2_{13}\chi_1 + c^2_{23}\chi_2 + c^2_{33}\chi_3, \qquad \ldots\ldots(3)$$

where χ_1, χ_2, χ_3 are the principal susceptibilities. It is worth noting that we could define the components of the magnetic field strength in terms of the components of the induced moment by an equation such as

$$H_i = \Omega_{ki}M_k,$$

where Ω_{ki} would be coefficients depending on the choice of the coordinate axes. χ_{ki} and Ω_{ki} would bear a similar relation to one another as conductivity coefficients bear to resistance coefficients. In practice, however, we only make use of χ's.

3. Limitation of the number of coefficients of susceptibility by the crystal symmetry.

Exactly the same considerations apply here as to the other second order tensors, so that reference should be made to Chapter I for the analysis, and to p. 8 for the table of non-zero coefficients in the various crystal systems.

4. Principles underlying experimental methods.

4.1. The couple exerted on a crystal by a uniform magnetic field.

It follows from paragraph 1 that the effect of a uniform magnetic field on a diamagnetic or paramagnetic crystal is independent of the shape of the crystal. We shall therefore consider the action on a rectangular bar of volume v which is free to rotate about an axis parallel to one edge. The magnetic field, H, and the induced moment, M, will not generally have the same direction (fig. 49). If the angle between them is ϕ then the couple, G, exerted on the crystal is given by

$$G = M \cdot H \sin\phi.$$

Both M and H, being vector quantities, may be resolved into components M_1, M_2, H_1, H_2 parallel to the directions of the

principal susceptibilities perpendicular to the axis of rotation
of the crystal. The resultant
couple G is equal to the difference of two couples, namely,
$M_1 H_2$ acting in an anticlock-
wise direction and $M_2 H_1$ acting
in a clockwise direction. Thus

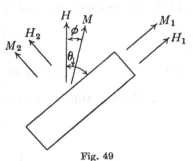

$$G = M_1 H_2 - M_2 H_1.$$

Now

$$H_1 = H \cos\theta, \quad H_2 = H \sin\theta.$$

Fig. 49

If the principal susceptibilities are χ_1 and χ_2, respectively, then

$$M_1 = v\chi_1 H_1, \quad M_2 = v\chi_2 H_2.$$

Hence

$$\begin{aligned} G &= v\chi_1 H_1 H_2 - v\chi_2 H_2 H_1 \\ &= vH^2 \sin\theta \cdot \cos\theta(\chi_1 - \chi_2) \\ &= \tfrac{1}{2}vH^2 \sin 2\theta(\chi_1 - \chi_2). \end{aligned}$$

4.2. The use of oscillations about a fixed axis perpendicular to a magnetic field to measure the differences between the principal susceptibilities of a diamagnetic or paramagnetic crystal.

As shown above, a crystal of any shape suspended in a uniform magnetic field experiences a couple equal to

$$\tfrac{1}{2}vH^2 \sin 2\theta(\chi_1 - \chi_2),$$

where v = volume of the crystal and θ = angle between χ_1 and H; and χ_1 and χ_2 are the principal susceptibilities in the plane perpendicular to the axis of rotation.

If θ is small this couple reduces to

$$G = vH^2\theta(\chi_1 - \chi_2).$$

The restoring couple per unit angular twist, k, is related to the time of swing, T_0, of a body of moment of inertia I, by the equation

$$k = \frac{4\pi^2 I}{T_0^2},$$

but in a magnetic field the restoring couple is changed so that if k' is the new restoring couple per unit angular twist,

$$k' = k + \frac{G}{\theta} = k + vH^2(\chi_1 - \chi_2)$$

and

$$\chi_1 - \chi_2 = \frac{1}{vH^2}(k' - k).$$

If T_1 is the new time of oscillation,

$$k' = \frac{4\pi^2 I}{T_1^2},$$

or

$$\chi_1 - \chi_2 = \frac{1}{vH^2} 4\pi^2 I \left(\frac{1}{T_1^2} - \frac{1}{T_0^2}\right).$$

If the value of I is difficult to obtain because the crystal is small, or irregularly shaped, a small glass rod of known moment of inertia is hung on the suspension and the time of oscillation found. The crystal is then attached and the time of oscillation redetermined, whence I may be found.

A modification of this method is probably more generally useful.† The crystal is mounted with one of its principal magnetic axes parallel to a very thin supporting fibre carried by a torsion head. In the magnetic field the crystal takes up a definite orientation and the torsion head is rotated until the crystal does not move on switching on the field. The setting of the torsion head is noted in this position. With the magnetic field on, the torsion head is turned continuously in one direction until the crystal has rotated through 45°. The couple is then a maximum and a slight further rotation of the torsion head causes the crystal to spin round through a large angle. If α_c is the angle of rotation corresponding to this critical setting, then

$$G = k\left(\alpha_c - \frac{\pi}{4}\right) = \tfrac{1}{2}vH^2(\chi_1 - \chi_2),$$

where k is the restoring force per unit angular displacement. Hence

$$\chi_1 - \chi_2 = \frac{2k(\alpha_c - \pi/4)}{vH^2}.$$

† K. S. Krishnan and S. Banerjee, *Philos. Trans.* A, **234**, 265 (1935).

k may be found by attaching to the fibre a body of known moment of inertia and finding the time of oscillation.

Observations of this kind can give only the difference between the principal susceptibilities lying in the plane perpendicular to the axis of oscillation. If the crystal is oscillated about the three principal axes of magnetic suscepti- bility the values of $\chi_1 - \chi_2$, $\chi_2 - \chi_3$, $\chi_3 - \chi_1$ may be obtained, but to determine the values of χ_1, χ_2 and χ_3, separately, further measurements must be made.

4.3. The measurement of absolute susceptibilities from the translation in a non-uniform magnetic field.

The general theory of the movement of a piece of magnetis- able matter in a magnetic field is already available in text-

books.† In the usual ex- perimental arrangement the crystal C (see fig. 50) is mounted in the space be- tween the poles N, S of an electromagnet and provision is made for the crystal to move only at right angles to the lines of force, i.e. along the dotted line in the figure. The susceptibility measured

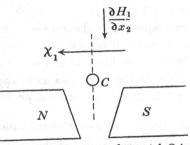

Fig. 50. The movement of a crystal, C, in an inhomogeneous magnetic field.

by the force exerted on C refers to a direction in the crystal denoted by the subscript 1, parallel to the magnetic field. The direction of movement is denoted by the subscript 2 and the observed force, F_2, is proportional to the component, H_1, of the field. It may be shown that

$$F_2 = v\chi_1 H_1 \frac{\partial H_1}{\partial x_2}. \qquad \ldots\ldots(4)$$

Another arrangement of pole pieces is sometimes used and

† F. B. Pidduck, *A Treatise on Electricity*, p. 234 (Camb. Univ. Press, 1916).

then the direction of movement is made to coincide with the direction of the field. The corresponding equation would be

$$F_1 = v\chi_1 H_1 \frac{\partial H_1}{\partial x_1}. \qquad \ldots\ldots(5)$$

It should be noted that equations (4), (5) are only applicable when one of the principal susceptibility axes is parallel to the magnetic field.

If the magnetic field is in turn made parallel to the directions corresponding to χ_1, χ_2, χ_3, we can obtain complete information concerning the magnetic susceptibilities of the crystal.

The mean value of χ may be obtained by using powder, and this value may be combined with the measurements of $\chi_1 - \chi_2$, etc., since $\chi_{\text{mean}} = (\chi_1 + \chi_2 + \chi_3)/3.$† In this way absolute values for the susceptibilities may be obtained. The method of obtaining the principal susceptibilities of a monoclinic crystal is similar to that used for finding the principal thermal expansions described on p. 26. The analytical treatment of the results is the same in both cases and will not be repeated here. It is also generally possible to find the directions of the principal axes approximately by trial and error and then to measure the principal susceptibilities directly.

5. Details of experimental methods.

Full descriptions of the available experimental methods are already to be found in text-books‡ and the following notes are only intended to supplement what is already published.

† The truth of this relation may be seen from the following considerations. Equation (3), p. 98, states that

$$\chi_{33}' = c_{13}{}^2\chi_1 + c_{23}{}^2\chi_2 + c_{33}{}^2\chi_3$$

and the mean value of χ can be obtained by integrating χ_{33}' over all possible directions. Such an integral would be equal to the sum of the integrals of each of the terms on the right-hand side of the equation taken over all directions. These integrals must be proportional respectively to χ_1, χ_2, χ_3, since there is no distinction between the integrals containing $c_{13}{}^2$, $c_{23}{}^2$, $c_{33}{}^2$ when all directions are considered. The factor of proportionality must be one-third, since

$$\chi_m = \frac{\chi_1 + \chi_2 + \chi_3}{3} \quad \text{when} \quad \chi_1 = \chi_2 = \chi_3.$$

‡ E. C. Stoner, *Magnetism and Atomic Structure*, p. 132 (Methuen, London, 1926).

Single crystals of metals are generally obtained as long rods
and the susceptibility in the direction of the length can be
obtained by the method of direct weighing (Gouy method).
The crystal is suspended from the beam of a balance so that
one end is in the centre of a strong magnetic field. Conical pole
pieces are arranged so that the field at the centre is horizontal.
If the field is of strength H at the centre, and not appreciable
at the other end of the crystal, then the force P acting on the
crystal (if its area of cross-section is A) is given by

$$P = \tfrac{1}{2}A\chi H^2.$$

Small crystals may be suspended in a non-uniform field by
a thin glass fibre and the movement of the crystal observed
with a microscope. The interpretation of the measurement is
simplest when the crystal is orientated so that a principal
axis is along the direction of the magnetic field. The great
difficulty in obtaining accurate values by this method is due
to the fact that the quantity $H(\partial H/\partial x)$ which determines the
force acting in the X-direction varies rapidly from point to
point. If the force required to displace the fibre a given distance can be found, then approximate results may be obtained
by this method. It can, however, be greatly improved by using
it in conjunction with the displacement principle.

A displacement principle can be applied to the measurement
of magnetic susceptibilities as well as to the measurement of
densities. The rotation or translation of a crystal in a magnetic
field depends, as we have seen in paragraph 1, both on the
susceptibility of the crystal and on that of the medium in
which it is immersed. If we wish to be strictly accurate in all
the equations we have so far considered, the susceptibility of
the crystal should be replaced by the difference between the
susceptibility of the crystal and the air or liquid in which it is
immersed. Thus if the crystal is immersed in a liquid which
has the same susceptibility per unit volume as the crystal in
the direction in which translation is possible, then there will
be no force acting on the crystal when a field is applied. This

principle makes it possible to overcome the difficulty of the rapid variation of $H(\partial H/\partial x)$, because if the crystal does not move from one position along the X-axis on applying the field it will not move from any point along this direction.

One practical method of making the susceptibility of the liquid the same as that of the crystal is to suspend it in a container fed by two tubes, one supplying a liquid with a higher and the other a liquid with a lower susceptibility than that of the crystal. When a mixture has been found in which the crystal does not move on applying the magnetic field, the liquid is removed and its density, refractive index or other determinative character found, so that its susceptibility may be obtained from a table.

A method applicable to *very small crystals* of certain substances has recently been devised.† It can only be used with certain substances because it is necessary to find a liquid mixture which has the same density and susceptibility as the crystal. These two conditions can be achieved over a certain range by mixing a diamagnetic with a paramagnetic solution, e.g. a solution of potassium bromide with one of nickel chloride. The experiment is carried out as follows. The crystals are floated in the solution contained in a vertical glass tube placed between the pole pieces of a magnet. These are so arranged that the magnetic field through the tube is everywhere horizontal and decreases rapidly from the bottom of the tube upwards. On applying the field the crystals first orientate themselves so that the maximum (algebraic) susceptibility is along the lines of force and then move bodily upwards or downwards according as the crystal has a smaller or greater susceptibility than the liquid. The crystals are observed by a telemicroscope mounted independently of the magnet. When the shape of the crystal and the orientation of its magnetic axes make it suitable for examination by this method, as, for instance, in di-benz-anthracene, crystals weighing no more than 2×10^{-5} g.m may be examined.

† K. S. Krishnan and S. Banerjee, *Z. Krist.* **91**, 173 (1935).

6. The correlation of the magnetic properties of paramagnetic and diamagnetic crystals with their crystal structures.

6.1. Introduction.

A number of authors† have determined theoretically the influence which the environment of certain ions exerts on their paramagnetic properties, but it is outside the scope of this book to consider this work. The gram-molecular susceptibilities for double sulphates of iron and ammonium, or nickel and ammonium, are about 11,000 and 4000 respectively, but the anisotropy is only about 2500 and 100 respectively. Small anisotropy and relatively large susceptibility is characteristic of many salts containing manganese, iron, nickel and cobalt. Much more is known about the relationship of magnetic properties to crystal structure in diamagnetic crystals than in paramagnetic. Both inorganic and organic substances have been investigated, but most of the latter are aromatic compounds. In the aromatic substances the molecules are well separated from each other, and the influence which one molecule exerts on the magnetic properties of a neighbouring molecule is small. We may therefore express the observed principal susceptibilities in terms of molecular constants in the way shown below. Thus we can go further than merely effect a correlation between the magnetic anisotropy and the structure, we can correlate it with the size and shape of the constituent molecules.

6.2. Magnetic anisotropy of inorganic crystals.

Using the division of crystal structures described in Chapter II, the following tables show the relations between magnetic anisotropy and crystal structure.

† J. H. van Vleck, *Electric and Magnetic Susceptibilities* (Oxford Univ. Press, 1932); N. F. Mott and H. Jones, *Properties of Metals and Alloys* (Oxford Univ. Press, 1936); A. H. Wilson, *The Theory of Metals* (Cambridge Univ. Press, 1936).

Isosthenic lattices.

The unit of $\chi_b - \chi_c$, etc. is the c.gm.-mol.s.e.m.u. These examples of isosthenic lattices have small magnetic anisotropies.

TABLE XVI. *Magnetic anisotropy of some rhombic sulphates*

Formula	Name	$\chi_b - \chi_c$ $(\times 10^6)$	$\chi_c - \chi_a$ $(\times 10^6)$
BaSO$_4$	Barytes†	0·72	0·62
SrSO$_4$	Celestite†	0·74	0·99
CaSO$_4$	Anhydrite†	0·25	−0·65

Layer lattices.

TABLE XVII. *Magnetic anisotropy of crystals with layer lattices*

Formula	Name	$\chi_\perp - \chi_\parallel$ $\chi_a - \chi_c$ $\chi_1 - \chi_2$ $(\times 10^6)$	$\chi_b - \chi_c$ $\chi_3 - \chi_2$ $(\times 10^6)$
Sb	Antimony‡	105	—
Bi	Bismuth§	−90	—
NaNO$_3$‖	Sodium nitrate†	4·9	—
CaCO$_3$‖	Calcite†	4·1	—
	Aragonite†	4·0	4·2
SrCO$_3$‖	Strontianite†	4·8	4·8
KNO$_3$‖	Potassium nitrate†	4·8	4·9

Antimony and bismuth are particularly interesting because the algebraic susceptibility of antimony is greatest perpendicular to the trigonal axis, while in bismuth, which has the same type of structure, it is parallel to the principal axis. The algebraic susceptibility for the rhombohedral and rhombic

† K. S. Krishnan, B. C. Guha and S. Banerjee, *Philos. Trans.* A, **231**, 235 (1933).
‡ D. Schoenberg and M. Z. Uddin, *Proc. Camb. Phil. Soc.* **32**, 499 (1936).
§ A. Goetz and A. B. Focke, *Phys. Rev.* **45**, 170 (1934).
‖ These crystals are included with the layer lattices because the existence of parallel planes of planar ions gives them many of the physical properties of layer lattices. They are not, however, true layer lattices, as is shown by the absence of a unique good cleavage perpendicular to the principal axis.

carbonates is greatest in a direction parallel to the plane of the CO_3 groups. Similarly, in the nitrates, the greatest susceptibility lies in a direction parallel to the NO_3 groups.

Three-dimensional framework structure.

The only example of this type of structure for which the diamagnetic data are available is quartz. $\chi_\perp - \chi_\parallel = 0.12 \times 10^{-6}$, a value which is small compared with the anisotropies of the carbonates and corresponds to the nature of the structure.

A hydrated paramagnetic salt.

The only detailed study combining the X-ray analysis of the crystal structure and the magnetic data of a paramagnetic substance relates to copper sulphate, $CuSO_4.5H_2O$.[†] Each copper ion is surrounded by four water molecules at a certain distance from it and two oxygens at a greater distance, so that together they form an octahedron which is drawn out along the line joining the two oxygens. There are two copper ions per unit cell and the planes of water molecules surrounding them are approximately perpendicular to one another. From a study of the crystal structure it is therefore to be expected that the magnetic properties would correspond to tetragonal symmetry, the planes of water molecules being parallel to mutually perpendicular planes of symmetry which intersect in a tetrad axis. This prediction is fulfilled, for it is found experimentally that the susceptibility has approximate uniaxial symmetry and that the direction of the unique axis coincides with the direction in which the planes of water molecules round the copper ions intersect.

6.3. Magnetic anisotropy of aromatic crystals.

A great wealth of results derived from aromatic compounds is available. A small selection is given in Table XVIII.

In the following paragraphs the method of deriving the molecular values from the principal crystal susceptibilities

---

† K. S. Krishnan and A. Mookherji, *Phys. Rev.* **50**, 860 (1936).

TABLE XVIII.† *Principal molecular susceptibilities of some aromatic hydrocarbons*

Formula	Name	Gram-molecular susceptibilities		
		$-K_1$ ($\times 10^6$)	$-K_2$ ($\times 10^6$)	$-K_3$ ($\times 10^6$)
	Di-benzyl	91	87	206
	Naphthalene	56	54	169
	Anthracene	76	63	252
	Terphenyl	97	88	271
	Quarterphenyl	122	110	372
	Chrysene	88	83	311

will be described. It is clear that the molecular susceptibilities K_1, K_2 and K_3 all increase with the size of the molecule and that K_3, corresponding to the direction perpendicular to the plane of the molecule, is much greater than the other two. It is also important to notice that K_1 is equal to K_2 within about 15 per cent, because use is made of this fact in the theoretical work which follows.

7.* The determination of the principal molecular susceptibilities of certain organic molecules.

It is possible in the triclinic, monoclinic and rhombic systems to correlate the principal susceptibilities of the crystal with those of the molecule. This is due to the fact that in these

† K. Lonsdale and K. S. Krishnan, *Proc. Roy. Soc.* A, **156**, 597 (1936).

systems, but in these only, there are few equivalent orientations of the molecule which are consistent with the symmetry. This makes it possible to determine the contribution of molecules which have the same orientation.

Since diamagnetic susceptibility is a centro-symmetric physical property it makes no difference to the calculation of molecular susceptibilities whether the crystal possesses a centre of symmetry or not. Thus we need only consider the holohedral classes of symmetry in the systems mentioned.

Triclinic system.

The measurement of the principal susceptibilities and the determination of their orientation with respect to the crystallographic axes is more difficult in this system than in any other. On the other hand, the relationship between the molecular orientation and the directions of the principal susceptibilities is simplest in this crystal system.

For the present purpose there is only one molecular orientation and consequently the principal susceptibility axes of molecule and crystal coincide, or

$$K_1 = \chi_1, \quad K_2 = \chi_2, \quad K_3 = \chi_3.$$

Monoclinic system.

If the direction-cosines of the susceptibility axes of one molecule relative to those of the crystal are given by the c_{ik}'s of the following scheme:

$$
\begin{array}{cccc}
 & \chi_1 & \chi_2 & \chi_3 \\
K_1 & c_{11} & c_{12} & c_{13} \\
K_2 & c_{21} & c_{22} & c_{23} \\
K_3 & c_{31} & c_{32} & c_{33}
\end{array}
\right\} , \qquad \ldots\ldots(6)
$$

then the corresponding direction-cosines for the molecule related by the plane of symmetry (010) are

$$
\left.
\begin{array}{ccc}
c_{11} & -c_{12} & c_{13} \\
c_{21} & -c_{22} & c_{23} \\
c_{31} & -c_{32} & c_{33}
\end{array}
\right\} . \qquad \ldots\ldots(7)
$$

These are the only two orientations which need be considered since the other two are obtained from these by the operation of a centre of symmetry.

If all the molecules are similar and similarly orientated the susceptibility in any direction may be obtained from equation (3), p. 98, namely,

$$\chi'_{33} = c_{13}^2 \chi_1 + c_{23}^2 \chi_2 + c_{33}^2 \chi_3,$$

by replacing χ_i by K_i.

To calculate the principal crystalline susceptibilities in terms of the principal molecular susceptibilities, we may proceed as follows. The susceptibility which the crystal would have in direction χ_1 if all the molecules were orientated as defined in equation (6) is given by

$$\chi_1 = c_{11}^2 K_1 + c_{21}^2 K_2 + c_{31}^2 K_3.$$

Half of the molecules have the orientation (7), but this does not alter the value of χ_1 since the squares of the direction-cosines appear in this formula. The three simultaneous equations defining χ_1, χ_2, χ_3 contain K_1, K_2, K_3 as unknowns and the latter can therefore be found by solving the equations. It is important to notice that the direction-cosines c_{ik} relating K_i and χ_k are found by assuming that the principal magnetic axes coincide with the length, breadth and thickness of the molecule. There is here one inherent limitation in the determination of the K's, for it is not always clear how the magnetic axes are related to the shape of the molecule. Thus for the flat aromatic hydrocarbons, naphthalene and anthracene, no ambiguity arises, because K_1 is parallel to the length of the molecule, K_3 perpendicular to its plane and K_2 is perpendicular to K_1 and K_3. For dibenzyl, $C_6H_5CH_2CH_2C_6H_5$, ambiguity does arise because the planes of the benzene rings, though parallel to one another, are not co-planar. In such a case special considerations must be applied which will not be discussed here.

Rhombic system.

The four molecular orientations possible in this system are given by the direction-cosine schemes:

	χ_1	χ_2	χ_3	χ_1	χ_2	χ_3	χ_1	χ_2	χ_3
K_1	c_{11}	c_{12}	c_{13}	$-c_{11}$	c_{12}	c_{13}	c_{11}	$-c_{12}$	c_{13}
K_2	c_{21}	c_{22}	c_{23}	$-c_{21}$	c_{22}	c_{23}	c_{21}	$-c_{22}$	c_{23}
K_3	c_{31}	c_{32}	c_{33}	$-c_{31}$	c_{32}	c_{33}	c_{31}	$-c_{32}$	c_{33}

$$c_{11} \quad c_{12} \quad -c_{13}$$
$$c_{21} \quad c_{22} \quad -c_{23}$$
$$c_{31} \quad c_{32} \quad -c_{33}$$

Again, it follows that since the relation between χ_i and K_k involves only the squares of the direction-cosines that all these sets of cosines lead to the same set of simultaneous equations. K_1, K_2, K_3 can therefore be found as in the monoclinic system.

Uniaxial and cubic systems.

In uniaxial systems the susceptibility of the crystal in all directions perpendicular to the principal axis is the same, hence $\chi_1 = \chi_2$. There are thus no longer three simultaneous equations from which K_1, K_2 and K_3 may be calculated. Similarly, in the cubic system $\chi_1 = \chi_2 = \chi_3$ and here also the magnetic constants of the molecule cannot be obtained.

8.* The calculation of molecular orientation from magnetic susceptibilities alone.

In the preceding paragraph we have seen how to obtain the principal molecular susceptibilities, being given the orientation of the molecules within the lattice framework, and also the principal susceptibilities of the crystal. The converse procedure of calculating the orientation of the molecule from its molecular susceptibilities and the measured susceptibilities of the crystal can be partially carried out. If two conditions are fulfilled, we may find the direction of the normal to the plane of the molecule. These two conditions are (1) that $K_1 = K_2$ and (2) the number of molecules per unit cell is not greater than the minimum required by the symmetry. The relations between χ_i and K_k become

$$\chi_1 = (c_{11}^2 + c_{21}^2) K_1 + c_{31}^2 K_3$$
$$= (1 - c_{31}^2) K_1 + c_{31}^2 K_3$$
$$= K_1 + c_{31}^2 (K_3 - K_1)$$

or
$$c_{31}^2 = \frac{\chi_1 - K_1}{K_3 - K_1},$$

or in general
$$c_{3i}^2 = \frac{\chi_i - K_1}{K_3 - K_1}.$$

Unless the second condition is fulfilled misleading results are obtained when the above calculation is carried through. For example, if in a monoclinic crystal there were four molecules per unit cell, though two were sufficient for the symmetry requirements, then the values of χ_i would be due to the combined action of the two differently orientated pairs of molecules, and c_{3i} calculated as above would not refer to either pair.

9. A theory of the molecular susceptibilities of aromatic compounds.

A quantum mechanical treatment[†] of the diamagnetic anisotropy of flat aromatic molecules leads to the result that the difference between the susceptibility parallel and perpendicular to the planes of the molecules can be obtained from the following model. If a wire framework, having the same form as the molecule, be placed perpendicular to a magnetic field which increases uniformly with time, the magnetic moment of the framework due to its induced current is proportional to the diamagnetic anisotropy of the molecule. The following examples illustrate the application of the method to naphthalene and anthracene. Fig. 51 shows the shape of the wire framework appropriate for naphthalene.

† L. Pauling, *J. Chem. Phys.* **4**, 673 (1936).

It is clear from the symmetry of the figure that under the influence of the magnetic field the points A and B will be at the same potential. Let the current flowing round the wires be i. Then the induced e.m.f. due to the changing flux is proportional to the area of the molecule. We take the area of one benzene ring to be six units of area.

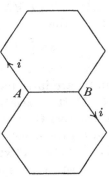

Then the induced e.m.f. $= 2 \times 6 = 12$. The resistances of the wires are proportional to their lengths, and if the unit of length is AB, the total resistance is 10. Thus

$$10i = 12, \quad \text{or} \quad i = \tfrac{6}{5}.$$

Fig. 51. Diagram illustrating the calculation of the magnetic anisotropy of the molecule of naphthalene.

Finally, the equivalent magnetic shell, k, has a pole strength equal to the area of the network† multiplied by the current, i.e.

$$2 \times \tfrac{6}{5} = \tfrac{12}{5}.$$

Anthracene.

The calculation of the currents in the parts of the network corresponding to anthracene is most simply done by supposing that cells, of voltage equal to the induced electromotive force, are introduced in each benzene ring, as shown in fig. 52, and then to apply Kirchhoff's laws to the net. Thus if the e.m.f. of the cell is 6, we have in the circuit $ABCDEF$

$$6 = 5i_3 - i_2,$$

in the circuit $IJEDGHI$

$$6 = 4i_1 + 2i_2,$$

and also

$$i_1 = i_2 + i_3.$$

Solving these equations we obtain

$$i_1 = \tfrac{24}{17}, \quad i_2 = \tfrac{3}{17}, \quad i_3 = \tfrac{21}{17}.$$

† The unit of area, in calculating the equivalent magnetic shell, is put equal to that of a whole benzene ring, because the factor -38×10^{-6} (*vide infra*) is obtained on this basis.

The resultant magnetic moment, k, is the sum of those due to a current of i_3 in the circuit $ABCDGHKLMNIJEF$ and a current of i_2 in the circuit $EDGHIJ$, i.e.

$$3 \times \tfrac{21}{17} + 1 \times \tfrac{3}{17} = \tfrac{66}{17}.$$

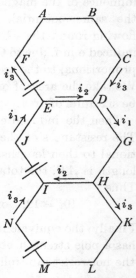

Two correction factors, both nearly unity, are introduced to take account of (1) the variation of the electric density per C—C bond in the molecule and (2) the effect of assuming a rectilinear network when the theory on which the work is based indicates that the electrons in the outer parts of the molecules have orbitals corresponding to arcs of circles passing through the points of the hexagons. In naphthalene there are 10 effective C—C bonds out of a total of 11 so that σ, the first factor, is $\tfrac{10}{11}$, whilst in anthracene, 14 out of 16 bonds carry most of the current, and $\sigma = \tfrac{7}{8}$. The second factor,

Fig. 52. Diagram illustrating the calculation of the magnetic anisotropy of the molecule of anthracene.

f, is a maximum for the small aromatic molecules, being 1·29 for benzene, C_6H_6, and 1·06 for $C_{54}H_{18}$.

If ΔK is the gram-molecular anisotropy, i.e. $(K_3 - K_1)$, the theory mentioned above gives the relation

$$\Delta K = -38 \cdot 0 \times 10^{-6} k\sigma f$$
$$= -104 \times 10^{-6} \text{ for naphthalene}$$
$$= -159 \times 10^{-6} \text{ for anthracene.}$$

It now remains to calculate K_1 and K_3 separately. The additivity of the susceptibilities of atoms in molecules has been well established (Pascal's law), and we may therefore derive the susceptibility parallel to the plane of the molecule by simple summation of the contributions of the constituent atoms. The gram-molecular susceptibility of H_2 is $-4 \cdot 0 \times 10^{-6}$, and we may therefore take the gram-molecular susceptibility of H bound to C as $-2 \cdot 0 \times 10^{-6}$. The gram-molecular suscepti-bility of C in diamond is $-6 \cdot 0 \times 10^{-6}$. In aliphatic carbon compounds each atom is bound to four neighbours, whereas in aromatic compounds there are only three carbons in the plane of the molecule and the fourth is considered in the cal-culation of ΔK. Thus for the aromatic carbon atoms the gram-molecular susceptibility is taken as $-4 \cdot 50 \times 10^{-6}$. Finally, we may write

$$K_1 = -(2 \cdot 0 n_H + 6 \cdot 0 n_{C.al} + 4 \cdot 5 n_{C.ar}) \times 10^{-6},$$

where n_H, $n_{c.al}$, $n_{c.ar}$ are the numbers of hydrogen, aliphatic carbon and aromatic carbon atoms per molecule respectively. It is assumed that the principal susceptibilities in the plane of the molecule (K_1 and K_2) are equal. The experimental results show that these constants are practically equal in most substances. K_3 may then be determined from the relation

$$K_3 = K_1 + \Delta K.$$

10. Comparison of theoretical and experimental values for the molecular susceptibilities of aromatic molecules.

Not many substances have been studied so carefully that one can work out reliable values of the principal molecular susceptibilities from the magnetic properties and molecular orientation. Some of those that have been so studied are included in Table XIX.

It will be seen that the agreement between the calculated and observed values is generally within 15 per cent. This difference is to be attributed partly to error of measurement, and partly to the simplifying assumptions made in the theory. The agreement is good enough, however, to show that the theory is well founded. Thanks to the small interaction between neighbouring molecules, the study of the magnetic properties of these organic substances has been carried nearer to com-pletion than the study of any other property shown by crystals.

TABLE XIX. *Comparison of the calculated and observed principal molecular susceptibilities of some aromatic compounds*

Name	Formula	$-K_1$ obs.	$-K_2$ obs.	$-K_1(K_2)$ calc.	K_3 Obs.	K_3 Calc.
Durene	$C_{10}H_{14}$	82	77	79	144	128
Hexamethylbenzene	$C_{12}H_{18}$	101	103	99	164	148
Terphenyl	$C_{18}H_{14}$	97	88	109	271	256
Quarterphenyl	$C_{24}H_{18}$	122	110	144	372	340
Naphthalene	$C_{10}H_8$	56	54	61	169	165
Anthracene	$C_{14}H_{10}$	76	63	83	252	242
Chrysene	$C_{18}H_{12}$	88	83	105	311	319

ELECTRIC INDUCTION

11. Introduction.

The earliest quantitative work on the behaviour of crystals in electric fields was carried out in the middle of last century, at about the same time as the corresponding magnetic investigations. The experimental methods employed were very similar. In one type of experiment the crystal was hung on a thread in a uniform electric field maintained between large parallel plates, and the orientation assumed by the crystal under the influence of the field was observed. In another experiment a diverging electric field was produced by maintaining a metal sphere of relatively small diameter at a high potential. The crystal being tested was supported on a torsion balance, or in some other way, so that the attraction exerted by the sphere could be measured. It was early observed that most crystals when suspended so as to be free to rotate in an electric field take up a certain orientation when the field is switched on, but then slowly change to another orientation. This is due to the fact that, unlike a magnetic pole, an electric charge can move slowly over the surface of an imperfect insulator, and this led to the introduction of other methods of experiment.

The difficulty introduced by the leakage of the electric charges was overcome by rapidly reversing the direction of the field, so that the reaction between the induced charge and the electric field occurred before the charge could leak away. Reference to the equation dealing with the torque exerted on a crystal by a uniform magnetic field, p. 99, will explain why reversing the field leaves the deflection unchanged. If the susceptibility is replaced by the polarisation k, i.e. the electric moment per unit volume induced by an electric field of unit strength, and H by E, the electric field strength, then the maximum couple G acting on a volume v of the crystal is

$$G = \tfrac{1}{2}vE^2(k_1 - k_2),$$

where k_1 and k_2 are the principal polarisation coefficients.

G depends on the square of E, and is therefore unaffected by reversing the direction of the field. Similarly, the attraction of a crystal towards a charged sphere along a direction Ox is determined by $E(\partial E/\partial x)$, which also remains unchanged when the field is reversed. The early attempts at reversing the field rapidly all depended on mechanical commutators, and the resulting frequency of commutation was not sufficiently high to be useful with many crystals. The discovery of Hertzian waves and, in particular, the application of Lecher wires made it possible to employ the very rapid oscillations which can be developed in electrical circuits.

The early measurements of the electric induction in crystals determined the induced electric moment per unit field strength. To avoid errors due to leakage of charge over and through the crystal, it was found better to use another experimental method, in which the dielectric constant was measured, i.e. the ratio of the capacity of a condenser with the crystal filling the space between the conducting plates, to the capacity of the same condenser with a vacuum between the plates. The relation between the electric polarisation, k (induced electric moment per unit volume per unit field), and the dielectric constant is given in some text-books,† but it is derived here by a slightly different method. Let the charge per unit area of the insulated plate of a parallel plate condenser with vacuum between the plates be q. Neglecting edge effects the electric field is $4\pi q$. If this field is kept constant when the crystal is placed between the plates the charge per unit area induced on the dielectric is $4\pi qk$. To keep the potential difference between the plates constant this charge must be added to the insulated plate because $4\pi qk$ units of charge have been effectively bound by the polarisation of the dielectric. Hence the total charge q' on the condenser is given by

$$q' = q + 4\pi qk.$$

† F. B. Pidduck, *A Treatise on Electricity*, p. 92 (Cambridge Univ. Press, 1916).

Since, by definition, the dielectric constant, ϵ, is given by

$$\epsilon = \frac{q'}{q},$$

$$\epsilon = 1 + 4\pi k.$$

12. Dielectric constants and crystal symmetry.

The general relations between the symmetry of the crystal and the dielectric constants ϵ_{ik} are just the same as apply to the magnetic susceptibilities χ_{ik} or any other second order tensor components. The proof that $\epsilon_{ik} = \epsilon_{ki}$ follows the same lines as that showing the equality of χ_{ki} and χ_{ik} (see Appendix II).

13. Experimental methods of measuring the dielectric constants of crystals.

Almost all modern methods of measuring the dielectric properties of crystals employ oscillatory circuits of the type used in radio technique.

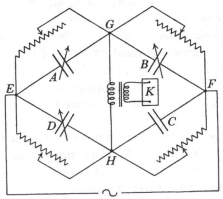

Fig. 53. Bridge circuit used for finding the dielectric constant of crystals.

One of the common arrangements is shown diagrammatically in fig. 53. It represents a capacity bridge in which A, B and D are standard variable condensers and C is the condenser between the plates of which the crystal may be placed.

Alternating current of the desired frequency is fed in at the points E and F and the bridge is said to be balanced when no current flows in the arm GH. The current in GH is detected by an amplifier K connected inductively to this arm. If the crystal is a perfect insulator and the capacities of the condensers be denoted c_a, c_b, c_c and c_d, then, when the bridge is balanced,

$$\frac{c_a}{c_b} = \frac{c_d}{c_c},$$

and since c_a, c_b and c_d can be read off from the calibrations of the condensers, c_c may readily be found. Usually it is necessary to make provision for the conductivity of the crystal. This is done by placing high variable non-inductive resistances in parallel with the condensers. Unless these resistances as well as the condensers are appropriately adjusted it is impossible to reduce the current in GH to zero. The bridge is balanced both with and without the crystal between the plates and the ratio of the capacities gives the dielectric constant. This method is suitable for frequencies of any value but requires three expensive standard condensers. Crystals may be used even when their linear dimensions are no more than a few millimetres. There are several variants of this bridge, one of which employs one standard condenser and two calibrated resistances, but apart from questions of cost and convenience in use there is no important difference between these various bridges.

Another method which is more often used is illustrated diagrammatically in fig. 54. The apparatus† consists of two high frequency oscillators A and B, one of which, say A, is provided with a variable condenser from the settings of which the frequency of A can be read off. Oscillator B has a condenser between the plates of which the crystal may be placed and in parallel with it a standard variable condenser. Oscillators A and B are coupled and their heterodyne note (of frequency equal to the difference in frequency of the two

† R. C. Evans, *Phil. Mag.* **24**, 70 (1937).

oscillators) is amplified by an audio frequency amplifier C. The note from C is compared by ear with that from an oscillator D emitting a note of constant audible frequency of, say, 1000 cycles/sec. The experiment may be carried out in the following way. The crystal is placed between the plates of its condenser, and the standard condenser in parallel with it is varied until the heterodyne note has the frequency of D. The reading C_1 of the standard condenser is then taken. The crystal

A	D
Oscillator	Fixed frequency generator (1000 cycles)

B	C
Oscillator	Amplifier of heterodyne note from oscillators A and B

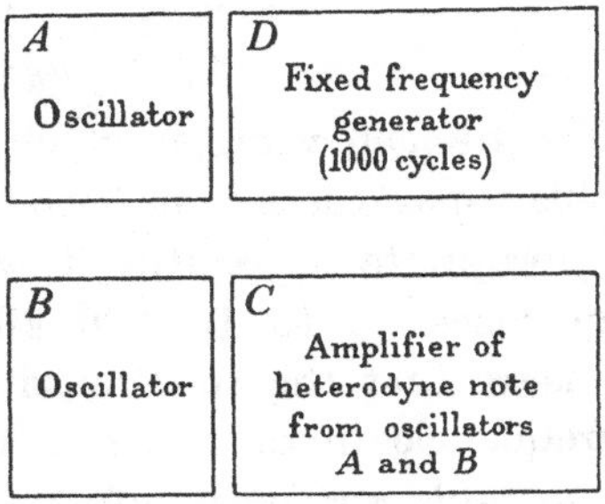

Fig. 54. Diagram indicating arrangement of circuits for obtaining a heterodyne note of given frequency.

is removed, leaving the plates of its condenser the same distance apart, and the standard condenser is reset (C_2) until the heterodyne note has the required frequency. The crystal condenser is disconnected from the standard condenser and the process repeated. If the last reading of the standard condenser attached to B is C_3, then the dielectric constant of the crystal, ϵ, is given by

$$\epsilon = \frac{C_3 - C_1}{C_3 - C_2}.$$

Usually an absolute measurement of this kind is not made, but the instrument is calibrated by using plates of known dielectric constants.

This method can be used to best advantage when the oscillators A and B are emitting high frequencies. This may be seen as follows. Suppose the ear can just detect a difference of frequency, Δ, at 1000 cycles/sec. and that the frequencies of oscillation of the currents in A and B are n_a and n_b respectively. The heterodyne frequency is then $n_a - n_b$, and if this can be just distinguished from the fixed frequency of D, then

$$n_a - n_b = 1000 \pm \Delta.$$

The error in finding n_b is fixed, being equal to $\pm \Delta$, but the percentage error decreases as n_b increases for

$$\frac{n_a - n_b}{n_a} = \frac{1000 \pm \Delta}{n_b + 1000 \pm \Delta} \approx \frac{1000}{n_b}.$$

The frequency range which can conveniently be covered by this apparatus is from 40 to 10,000 kilocycles/sec., though with appropriate apparatus it would be possible to obtain frequencies up to 100,000 kilocycles/sec. It is necessary to ensure that the conductivity of the crystal is small at the frequencies used, as otherwise the measured values of dielectric constants are inaccurate.

The nature and form of the electrodes present difficulties with the small crystals usually available. Mercury electrodes and thin metal films obtained by sputtering or evaporation in vacuum are often used. The form of the electrodes is frequently such that the end corrections are serious and the apparatus must be calibrated with identical electrodes placed on crystals of different known dielectric constants.

For isotropic powdered crystals it is often more convenient to use a displacement method than to find the dielectric constant by absolute measurements. If the condenser plates are immersed in a liquid of the same dielectric constant as the crystal, then, on introducing the latter between the plates of the condenser, the capacity of the condenser will be unchanged. This method may, of course, be applied to single anisotropic crystals as well as to isotropic powders.

14. The relation between the dielectric properties of crystals and their crystal structures.

The data available for correlating the dielectric properties of crystals with their crystal structures do not permit more than general relations to be derived. The closeness of packing of ions appears to be related to the magnitude of the dielectric constant. Thus among the halides of lithium and sodium (see Table XX) the highest dielectric constants occur in those compounds for which the ratio of the radii of kation to anion is most nearly that which corresponds to close-packing of the anion, i.e. 0·43. In the following table the radius ratios are given in brackets beside the dielectric constants.

TABLE XX

Dielectric constants and radius ratios of the alkali halides†

	F	Cl	Br	I
Li	9·2 (0·59)	11·05 (0·43)	12·1 (0·40)	11·03 (0·35)
Na	4·9 (0·74)	5·77 (0·54)	5·99 (0·50)	6·60 (0·44)
K	6·05 (1·00)	4·76 (0·73)	4·78 (0·68)	4·94 (0·60)
Rb	5·91 (1·12)	5·20 (0·82)	4·70 (0·76)	4·81 (0·68)

It will be seen that the lithium halides have higher dielectric constants than any others and that those which have radius ratios of 0·43 and less have about the same dielectric constants. In the second row the dielectric constants of the sodium halides increase as the radius ratios approach the close-packing value. It will be clear from the difference between the dielectric constants of lithium chloride and sodium iodide, which have nearly the same value of the radius ratio, that the radius ratio is by no means the only factor determining the dielectric constant. The radius ratios for the halides of potassium and rubidium are so far from 0·43 that these considerations could hardly be expected to apply. The three polymorphic forms of titanium dioxide afford another example of the relation between the dielectric constant and the closeness of packing. The following table compares the density and mean dielectric constants of the three crystals.

TABLE XXI

	Rutile	Brookite	Anatase
Density	4·21	4·11	3·87
Mean ϵ	114	78	48

† J. Errera, *Zeit. f. Elektrochemie*, **36**, 818, 1930.

Applying this observation to anisotropic crystals it is to be expected that the directions in which the ions are most closely packed will correspond to the direction of greatest dielectric constant. This expectation is to some extent fulfilled. Thus, in the rhombohedral carbonates, calcite $CaCO_3$, dolomite $CaMg(CO_3)_2$, rhodochrosite $MnCO_3$, siderite $FeCO_3$ and smithsonite $ZnCO_3$, the value of the dielectric constants parallel and perpendicular to the trigonal axis, i.e. $\epsilon_\parallel$ and $\epsilon_\perp$, are in the ratios of 0·89, 0·87, 0·86, 0·87, 0·99; magnesite, $MgCO_3$, is an exception because the ratio is 1·17.† The planes of the CO_3 ions are all perpendicular to the trigonal axes in these crystals and the oxygen ions are therefore most closely packed in directions perpendicular to the trigonal axis. The same general result holds for the rhombic carbonates, for in every case the least dielectric constant is observed when the electric field is applied parallel to the c axis, i.e. perpendicular to the planes of the CO_3 groups. The orthorhombic sulphates celestine $SrSO_4$, barytes $BaSO_4$ and anglesite $PbSO_4$ have a much greater dielectric constant parallel to the b axis than in the other principal directions (see Table XXII) and this is probably connected with the occurrence in the structure of lines of oxygen ions parallel to this direction.

TABLE XXII.† *Dielectric constants of some rhombic sulphates*

	ϵ_a	ϵ_b	ϵ_c
$SrSO_4$ (celestite)	7·7	18·5	8·3
$BaSO_4$ (barytes)	7·65	12·6	7·7
$PbSO_4$ (anglesite)	27·5	54·6	27·3

It is not difficult to explain qualitatively the commonly occurring relation between closeness of packing of ions and the magnitude of the dielectric constant. If an ion is isolated it will, under a given field, acquire a certain electric moment. If a second ion is placed beside it in the direction of the electric

† J. Errera und H. Brasseur, *Phys. Zeit.* **34**, 368 (1933).

field the induced moment of the second ion increases the strength of the field acting on the first. A string of ions in the direction of the field would thus tend to increase considerably the moment induced in any one ion. On the other hand, if a second ion were placed beside the first, in a direction perpendicular to the field, the induced moment of the second would reduce the field acting on the first and hence reduce its moment. Thus any arrangement of ions in strings or layers will tend to increase the dielectric constant in the direction of closest packing and reduce it in others (see also Chapter V, p. 183).

15. The dielectric properties of Rochelle salt, $NaKC_4H_4O_6 . 4H_2O$.

This crystal has the most remarkable dielectric properties of any known substance.† The dielectric constant in the direction of the a axis is abnormal, being dependent to a marked extent on the temperature, the strength and the frequency of the electric field applied, and the perfection of the crystal. The dielectric constants parallel to the b and c axes are about 60 and almost independent of temperature and the characteristics of the applied field. The temperature range over which the great dielectric constants can be observed is from about $-10°$ C. to $+20°$ C. The values of the dielectric constant ϵ_a below $-10°$ C. are about 150 and above $20°$ C. about 400: in the region of $40°$ C. the crystal becomes conducting. Various values up to 60,000 have been found for ϵ_a: the maximum value is obtained at a field strength of about 100 volt/cm. and above this value the dielectric constant decreases rapidly while the polarisation remains constant. This is analogous to the saturation obtained on magnetising ferromagnetic substances. Hysteresis, which is so characteristic a feature of the magnetisation of iron, is also shown by Rochelle salt when the field exceeds 50 volt/cm. All the effects

† A good review can be found in Hans Staub, *Naturwissenschaften*, **23**, 728 (1935).

depend on the purity and perfection of the crystal and to some extent on its previous electrical history. The change of dielectric constant with frequency of the applied field is equally remarkable.[†] There is a fairly uniform decrease from 62,000 at 30 cycles/sec. to 220 at 10^7 cycles/sec., and then the dielectric constant changes abruptly to negative values at 14×10^7 cycles/sec.[‡] This frequency appears to be a true constant of the material and not a value depending on arbitrary factors such as the size, shape or nature of the electrodes. The variation with chemical nature has been studied[§] by substituting NH_4^+ for K^+ and the dielectric constant falls to normal values with increase in the percentage of the ammonium salt in the mixed crystal. Up to the present no complete theory has been advanced to explain these facts, and until the crystal structure of this substance is found it is improbable that one can be given. There is, however, so close a resemblance between the magnetic effects in ferromagnetic substances and these electrical phenomena that the fundamental character of both must be similar. In general terms a theory of the dielectric properties of Rochelle salt is as follows. There exist in Rochelle salt between $-10°$ C. and $20°$ C. small regions in which the dipoles are all orientated in the same direction under the influence of their mutual fields, i.e. these regions are spontaneously polarised. The directions of electrification of the various regions are orientated at random, but when an electric field is applied first one and then another switch over into parallelism with the applied field. This accounts for the increase of the polarisation with strength of the electric field and the saturation which occurs when all the small regions are similarly orientated. The hysteresis arises from the influence of the small regions on one another, which tends to prevent them from losing their orientation on weakening the field.

† A. Zeleny and J. Valasek, *Phys. Rev.* (2), **46**, 450 (1934).

‡ This is the only recorded example of a negative dielectric constant which is independent of resonance phenomena due to the alternations of the electric field.

§ R. C. Evans, *Phil. Mag.* **24**, 70 (1937).

This theory is too general to account for the existence of small regions of spontaneous electrification and for the effects which depend on frequency, temperature, mechanical deformation and chemical composition. From analogy with other substances containing water of crystallisation and with ice, it is believed that the water molecules are responsible for these remarkable dielectric effects. In Chapter VI reference will be made to the piezo-electric properties of Rochelle salt, which are very intimately connected with the dielectric properties.

CHAPTER V

SOME PROBLEMS IN CRYSTAL OPTICS

The theory of Crystal Optics is dealt with to some extent in text-books on Optics and more fully in crystallographic, mineralogical and petrological works.† In this book, therefore, the whole of Crystal Optics will not be dealt with but only certain parts which are inadequately or incorrectly treated in the existing books.

1. Problems connected with the geometry of the indicatrix.

1.1. Geometrical properties of a triaxial ellipsoid.

There are three properties which we shall use in the course of this work:

(a) Every plane which passes through the centre of an ellipsoid cuts it in an ellipse.‡

(b) Tangent planes of the ellipsoid drawn at all points where it is cut by a central section intersect in lines which are parallel to the conjugate diameter of the central section.§ These lines in which the tangent planes intersect form a tangent cylinder.

† A. N. Winchell, *Elements of Optical Mineralogy*. Part I (Wiley and Sons, New York, 1937).

F. Pockels, *Lehrbuch der Kristalloptik* (Teubner, Leipzig and Berlin, 1906).

H. Rosenbusch and E. A. Wülfing, *Mikroskopische Physiographie der petrographisch wichtigen Mineralien*. Band I (E. Schweizerbartsche Verlagshandlung, 1904).

A. Johannsen, *Manual of Petrographic Methods* (McGraw-Hill, New York and London, 1918).

N. H. Hartshorne and A. Stuart, *Crystals and the Polarising Microscope* (Arnold, London, 1934).

L. Fletcher, "The Optical Indicatrix and the Transmission of Light in Crystals", *Min. Mag.* No. 44 (1891). (This can be obtained from the Mineralogical Society, London.)

G. Szivessy, *Handbuch der Physik*, vol. xx (Springer, Berlin, 1928).

‡ R. J. T. Bell, *Coordinate Geometry of Three Dimensions*, p. 107 (Macmillan, London, 1912).

§ *Ibid.* p. 110.

(c) The major axis OA (fig. 56) of any central section and its normal OP define a plane containing the normal AQ to the ellipsoid at the point where the major axis cuts it. A similar relation holds for the minor axis. The proof of this may be

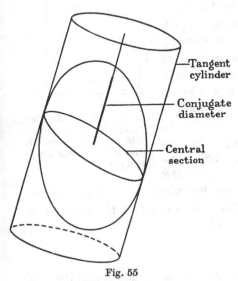

Fig. 55

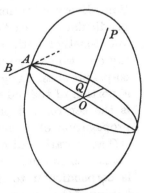

Fig. 56. Diagram showing that the normal OP to a section of the indicatrix is co-planar with the major axis OA of the section and the normal to the indicatrix at A.

given as follows. The tangent AB at A to the central section is a line perpendicular to OA and lies in the tangent plane to the ellipsoid at A. Hence the normal, AQ, to the tangent plane must be perpendicular to AB and therefore lie in the plane AOP which is perpendicular to plane OAB. Hence AQ (fig. 56) intersects OP.

1.2. The representation of the optical properties of a biaxial crystal by a triaxial ellipsoid.

No attempt will be made here to justify the validity of the constructions enunciated below. That the optical properties of crystals can be represented by the ellipsoid known as the indicatrix has been experimentally verified with a high degree of accuracy. The principal axes of the indicatrix are of lengths proportional to α, β and γ and are drawn parallel to the vibration directions of the light having the corresponding refractive indices. Any radius vector defines by its direction a vibration direction and its length is proportional to the refractive index of the light vibrating parallel to this direction. Every vibration direction is perpendicular to at least one wave-normal, and together with the corresponding ray direction they form a set of three co-planar lines. In fig. 57 AB represents a portion of the indicatrix of which O is the centre. OW is the direction of propagation of waves of light having AO as vibration direction. AN is the normal to the indicatrix at A and OR is perpendicular to AN. The direction of travel of the ray having AO as vibration direction is OR and the velocity of this ray is proportional to $1/AN$.

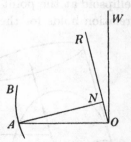

Fig. 57. The relation between the wave-normal OW, the ray direction OR and the vibration direction OA.

1.3. Wave-normals and rays.

The following construction enables us to determine the two rays correlated with any given wave-normal.

OW (fig. 58) is the direction of any wave-normal and the ellipse represents a section of the indicatrix perpendicular to OW. The major and minor axes are OA, OB and AQ, BS are normals to the indicatrix at A and B respectively. The rays associated with OW are OR_1, OR_2, which are perpendicular to AQ and BS respectively. It follows from property (c) of a triaxial ellipsoid described in paragraph 1 that OA, AQ, OR_1 and OW are all co-planar. This is also true of OB, BS, OR_2 and OW. It should be noted that the conjugate diameter to the section OAB does not coincide with either OR_1 or OR_2.

The converse problem of finding the two wave-normals correlated with a given ray direction may be solved in a similar manner. We must find two mutually perpendicular normals to the ellipsoid which are also perpendicular to the ray. The

locus of all points on the ellipsoid for which the normals are
perpendicular to the ray OR (fig. 59) is the ellipse DE in which
the ellipsoid is touched by the tangent cylinder parallel to

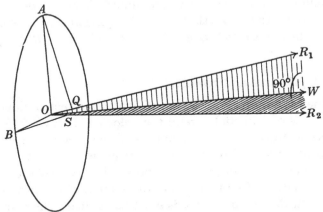

Fig. 58. The relation between a wave-normal OW and the
associated vibration directions OA, OB and rays OR_1, OR_2.

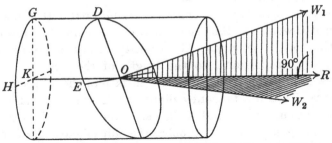

Fig. 59. The relation between a ray OR and the associated
vibration directions OD, OE and wave-normals OW_1, OW_2.

OR. The normals at all except four points on the ellipse DE
do not intersect OR. These four points are not, in general, the
ends of the major and minor axes of the ellipse DE, though
they are near to them.† Consider the right section GH of the

† It is necessary to emphasize this fact because an oft-quoted paper, G. Tunell,
"The ray-surface, the optical indicatrix, and their interrelation: An elementary
presentation for petrographers", J. Wash. Acad. Sci. 23, 325 (1933), is in error
about this. Following him, Roger and Kerr, Thin-section Mineralogy (McGraw-
Hill, New York and London, 1933), is also in error. A correct statement was given
in Fletcher's book and also in Pockel's.

tangent cylinder. The major and minor axes of this section pass through its centre K. Hence if DO is co-planar with GK, and EO with HK, the points D and E, where the normals to the ellipsoid intersect OR, are found. The wave-normals associated with OR are therefore OW_1, lying in the plane DOR perpendicular to DO, and OW_2 in the plane EOR, perpendicular to EO.

1.4. Optic axes and biradials.

When the normals to waves of light are directed along an optic axis of a biaxial crystal the velocity of travel is found experimentally to be independent of the inclination of the plane of vibration to the optic axial plane. Hence the radii of the indicatrix which correspond to these waves are all of equal length and co-planar, i.e. the section of the indicatrix perpendicular to an optic axis must be circular. There are, of course, two circular sections of a triaxial ellipsoid corresponding to the two optic axes.

Rays of light travel along an optic biradial with the same velocity, whatever the orientation of the plane of vibration to the optic axial plane. The velocity of ray propagation is inversely proportional to the length of the normal to the ellipsoid intercepted between the surface of the ellipsoid and the ray. Thus if a tangent cylinder be drawn to the ellipsoid parallel to an optic biradial the perpendiculars to the biradial from the points where it touches the ellipsoid are all of equal length. The right section of this tangent cylinder is therefore circular (fig. 55), and from the symmetry of the ellipsoid the right sections corresponding to both optic biradials must be of the same diameter.

1.5. Derivation of the ray surface† from the indicatrix.

Consider first the principal sections of the indicatrix; they are ellipses having axes proportional to the principal refractive indices. To each principal axis of this ellipse there corresponds

† For definitions of ray and wave surfaces see Appendix III.

an infinite number of rays lying in a plane perpendicular to this axis. The ray associated with a vibration direction which is nearly parallel to, say, the axis OX_1 is a line nearly parallel to the $X_2 OX_3$ plane (fig. 60). The directions of the rays corresponding to a small circle on the indicatrix round the point where OX_1 passes through it are all nearly parallel to the plane $X_2 OX_3$, and in the limit when the small circle becomes reduced to a point, the rays coincide with this plane. Thus the section of the ray surface defined by the plane passing through X_2 and X_3 contains a circle of radius proportional to $1/\alpha$. Similarly, the sections of the ray surface defined by planes containing $X_1 X_3, X_1 X_2$ contain circles of radii proportional to $1/\beta$ and $1/\gamma$ respectively.

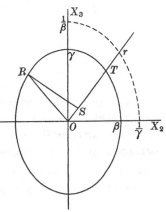

Fig. 60. The relation between a principal section of the indicatrix and the parallel principal section of the ray surface.

The principal planes of the ray surface have ellipses in addition to the circles. Consider the section of the indicatrix by a plane containing $X_2 X_3$. Any radius vector OR lying in the plane $X_2 OX_3$ has one corresponding ray, Or, which also lies in the plane. Or is perpendicular to the normal RS to the ellipse at R. Since Or and OR are conjugate radii,

$$RS \cdot OT = \beta\gamma.\dagger$$

Further, since the velocity of the ray Or is inversely proportional to RS, and the factor of proportionality may be chosen equal to unity, $Or = 1/RS$, hence $Or = OT/\beta\gamma$. Thus the locus of r is an ellipse similar to, and orientated in the same way as, the section $\beta O\gamma$ of the indicatrix. The principal axes of this section of the ray surface have axes proportional to $1/\beta$ and $1/\gamma$. In a similar way we may derive the form of the other

† S. L. Loney, *Coordinate Geometry*, p. 256 (Macmillan, London, 1920).

principal sections of the ray surface, and the results may be summarised as in Table XXIII.

TABLE XXIII. *Principal sections of the ray surface*

Plane of section	Radius of circle	Principal axes of ellipse	
X_2OX_3	$1/\alpha$	$1/\beta$	$1/\gamma$
X_1OX_3	$1/\beta$	$1/\gamma$	$1/\alpha$
X_1OX_2	$1/\gamma$	$1/\alpha$	$1/\beta$

1.6. The two planes which intersect in a wave-normal and each contain an optic axis meet at an angle which is internally and externally bisected by the vibration directions corresponding to the wave-normal.

Fig. 61 is a stereographic projection in which Q is the projection of a wave-normal and A_1, A_2 the projections of the

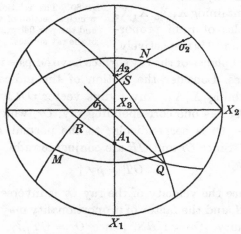

Fig. 61. Stereographic projection showing the wave-normal Q, and the associated vibration directions σ_1, σ_2. A_1, A_2 are the optic axes.

optic axes. The plane of the wave is represented by a great circle MN of which Q is the pole. This plane intersects the indicatrix in an ellipse the major and minor axes of which we suppose to be σ_1 and σ_2. The circular sections of the indicatrix are perpendicular to the optic axes and are represented by the circles X_2N, X_2M. Since the circular sections have the same radii the lines in which they intersect the ellipse represented by NM must be of equal length. Now the symmetry of an ellipse requires that the radii of equal length shall be equally inclined to the principal axes. Thus σ_1 must bisect the angle NM, and σ_2 the angle $N\overline{M}$. Since N is at the point of intersection of two great circles representing planes perpendicular to A_1 and Q respectively, N must represent a line perpendicular to the plane A_1Q. Similarly, M must represent a line perpendicular to the plane A_2Q. Since $M\sigma_1 = \sigma_1N$ and $NR = MS = 90°$, $R\sigma_1 = \sigma_1S$, i.e. the plane $Q\sigma_1$ bisects the angle A_1QA_2 internally, and σ_2Q bisects it externally.

1.7.* **The two planes which intersect in a ray direction and each contain an optic biradial meet at an angle which is internally and externally bisected by the vibration directions corresponding to the ray direction.**

Fig. 61 may be used for this paragraph if the poles and circles be supposed moved a little from their former positions. Q now represents a ray direction, A_1, A_2 optic biradials. The plane represented by MN is the conjugate plane to Q and the planes MX_2, NX_2 are taken as conjugate to A_2 and A_1 respectively. M, since it represents a direction common to the conjugate planes of Q and A_2, is the conjugate diameter of the plane QA_2. In fig. 62 the line T_2 represents the tangent plane to the indicatrix at the point where the line through M (fig. 61) cuts it and is therefore parallel to the plane containing the ray and the optic biradial A_2. Similarly, the plane, T_1, which is tangential to the indicatrix at the point where the line through N cuts it, is parallel to the plane containing the ray and the optic biradial A_1. In fig. 62 the ray OQ is supposed to be perpendicular to the paper, and the optic biradials, OA_1, OA_2, are drawn so as to show that they are inclined at different angles to the plane of the paper. The planes

A_1OQ, A_2OQ have been shown to be parallel respectively to
T_1 and T_2. Since T_1 is a tangent
to the indicatrix and parallel to
OA_1, the perpendicular distance
between them is equal to the
radius of the cylindrical tangent
cylinders parallel to the optic
biradials. Similarly, the perpendicular distance between T_2 and
OA_2 is equal to the same radius.
Finally, T_1 and T_2 are tangents to
the elliptical tangent cylinder
parallel to OQ. These conditions
can only be satisfied if T_1 and T_2,
and therefore the planes A_1OQ,
A_2OQ, are symmetrical with respect to the principal axes of the
right section of this tangent
cylinder, i.e. the vibration directions corresponding to OQ bisect
the angle between the planes
A_1OQ, A_2OQ.

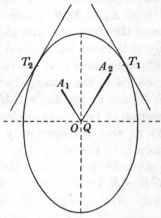

Fig. 62. Projection on a plane perpendicular to the ray OQ. T_1, T_2
are tangents to the indicatrix and
both are parallel to OQ. OA_1, OA_2
are the optic biradials to which
T_1 and T_2 are also respectively
parallel. The dotted lines indicate
the planes containing the vibration directions associated with
OQ.

1.8. Relation between the wave-normal direction, ray directions, and vibration directions. (Sylvester's construction.)

The results of paragraphs 1.6 and 1.7 can be combined into a
single construction, because the correlated vibration direction,
ray direction and wave-normal must all be co-planar. In fig. 63
the point Q represents the direction of a wave-normal, A_1, A_2
the optic axes, B_1, B_2 the optic biradials. The construction for
finding the direction of one of the rays S_1 corresponding to Q
is as follows. The great circles A_1Q, A_2Q meet at an angle which
is internally bisected by TQ and externally bisected by PQ.
The great circle TB_2PN is drawn to cut PQ at right angles and

N is fixed so that the angle $PN = B_2P$. The great circle NB_1 is then drawn to cut PQ in S_1. S_1 gives the direction of the ray associated with Q which has a vibration direction lying in the plane QS_1. From the right-angled triangles B_2PS_1, NPS_1 it is obvious that PQS_1 bisects the angle B_2S_1N. By construction PQS_1 bisects the angle A_2QA_1 and hence the ray S_1, wave-normal Q_1 and vibration direction, satisfy the requirements established in paragraphs 1.6 and 1.7. The construction for finding the second ray associated with Q is carried

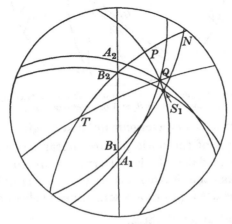

Fig. 63. Stereographic projection showing the optic axes A_1, A_2, optic biradials B_1, B_2, wave-normal Q and one corresponding ray S_1.

out in a similar way by drawing a great circle through B_2 to cut circle TQ at right angles, and then proceeding as before.

In fig. 64 the method of finding a wave-normal Q correlated with a given ray direction S_2 is illustrated. The angle between the great circles B_2S_2, B_1S_2 is bisected internally by the circle VS_2. A great circle A_2U is drawn so that the arc A_2U is perpendicularly bisected by the great circle VS_2. The great circle A_1U is drawn to cut VS_2 in Q. Q is then one of the directions of the wave-normals associated with S_2. The other may be found by drawing a circle through A_2 at right angles to the circle WS_2, the other bisector of the angle $B_1S_2B_2$, and proceeding as before.

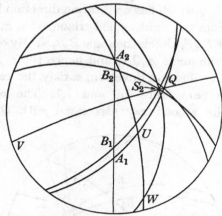

Fig. 64. Stereographic projection showing the optic axes A_1, A_2, optic biradials B_1, B_2, ray direction S_2 and one corresponding wave-normal Q.

2. Observations in convergent polarised light.

Descriptions of the optic pictures obtained in convergent light with uniaxial and biaxial crystals are to be found in several books, and further discussion here is only justified because these descriptions are sometimes misleading.

2.1. Uniaxial optic pictures.

The usual explanation of the formation of optic pictures in monochromatic light is illustrated in fig. 65. Two rays having outside the crystal the same inclination with respect to the optic axis enter the crystal and are resolved into ordinary and extraordinary rays which are transmitted in slightly different directions. An ordinary and an extraordinary ray both emerging

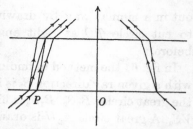

Fig. 65. The passage of light through a doubly refracting plate when the angle of incidence is not zero.

at the same point have vibration directions which are

mutually perpendicular. On emergence therefore, they do not interfere, but after transmission through the analyser, which is crossed with respect to the polariser, no light is observed if the path difference between the rays emerging at the same point is an integral multiple of λ. On the other hand, the light has a maximum intensity if the path difference is $(2n+1)\lambda/2$, where n is any integer. Up to this point the explanation though not entirely correct may be allowed. Several books continue the development of the discussion by saying that if

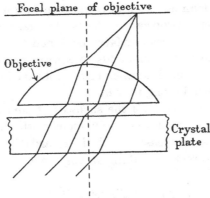

Fig. 66. The convergence to a point in the optic picture of a broad beam of light passing through the crystal plate.

O is the intersection of the axis (fig. 65) of the optical system with the crystal plate, as OP increases the corresponding rays become more and more inclined to the optic axis, and because the path difference between the ordinary and extraordinary rays continually increases, they give rise to alternate light and dark circles in the image. This presentation is false. The distance PO has practically no connection with the radius of the corresponding ring in the optic picture. The optical system used for observing optic pictures is a telescope of wide angle and, except for effects due to aberration of the lenses, a parallel beam will be brought to a focus at a single point in the focal plane of the objective. Fig. 66 shows a beam of rays lying in a principal section of the crystal and converging to a focus at a point on one of the rings of the picture. It is important to avoid the mistake mentioned above, because it is impossible to understand the formation of an image located at a definite point in the barrel of the polarising microscope, as long as each point in the optic picture is supposed to correspond to a particular ray.

2.2. Explanation of the characteristics of optic pictures.

The simplest way of deriving an explanation of the shapes of the circles, lemniscates and ovals seen in optic pictures is by constructing appropriate surfaces of equal retardation. If any point on one of these surfaces be joined to the origin, the direction of the line corresponds to a wave-normal direction and its length to the difference in the retardation of the two beams transmitted along that line.

(a) *For uniaxial crystals* the surfaces of equal retardation are cones having their axes coincident with the optic axis and semi-angles ϕ increasing with the retardation. The double refraction in a direction making an angle ϕ with the optic axis is given by

$$\mu' - \mu'' = (\mu_e - \mu_o)\sin^2\phi, \dagger$$

where μ_e, μ_o are the extraordinary and ordinary refractive indices. The path difference introduced between the ordinary and extraordinary ray in passing through the thickness t is $n\lambda$, where

$$n\lambda = t(\mu' - \mu'') = t(\mu_e - \mu_o)\sin^2\phi.$$

Thus if t is constant and ϕ_n is the value of ϕ corresponding to a path difference of $n\lambda$,

$$\sin^2\phi_n = \frac{n\lambda}{t(\mu_e - \mu_o)}.$$

If ϕ is small so that we may put $\sin\phi = \phi$,

ϕ_n is proportional to $\sqrt{n}$.

† This is a special case of the formula for biaxial crystals, p. 141.

When a uniform plate of crystal is cut with its faces perpendicular to the optic axis, and observed in convergent polarised light between crossed nicols, a series of concentric rings is seen. We may regard the rings as occurring where the cones of equal retardation cut the surface of the crystal, as indicated in fig. 67. In practice the values of ϕ will not increase as rapidly as $\sqrt{n}$ because t increases with the inclination of the rays to the optic axis. (Fig. 67 does not indicate that the light responsible for producing a given ring in the picture originated in a single point on the underside of the block.)

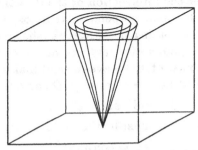

Fig. 67. The cones of equal retardation in a uniaxial crystal. The successive cones correspond to retardations of λ, 2λ, 3λ, etc.

(b) *In biaxial crystals* the surfaces of equal retardation have a rather complicated shape. We shall investigate their shape by finding the curves in which they intersect particular planes.

(1) Section of the surface of equal retardation by a plane perpendicular to the acute bisectrix at a distance d from the origin.

We shall assume in order to simplify the analysis that the optic axial angle $2V$ is small and that only directions near to the acute bisectrix are to be considered, so that $\sin \phi$ may be put equal to ϕ. The equation giving the retardation is then

$$n\lambda = t(\mu' - \mu'') = t(\alpha_3 - \alpha_1) \sin \phi \sin \phi', \dagger$$

† T. Preston, *The Theory of Light*, p. 357 (Macmillan, London, 1912); A. Johannsen, *Manual of Petrographic Methods*, p. 351 (McGraw-Hill, New York and London, 1918).

where α_3, α_1 are the greatest and least principal refractive indices and ϕ and ϕ' are the angles made by the direction under consideration with the directions of the optic axes. Using the approximation mentioned above,

$$n\lambda = t(\alpha_3 - \alpha_1)\,\phi\phi',$$

we may assume that t is constant for light travelling in the region not far removed from the optic axes, so that

$$n\lambda = k\phi\phi',$$

where k is a constant.

If we consider the intersection of the directions defined by ϕ and ϕ' with the plane perpendicular to Bx_a at a distance d from the origin we obtain a plane figure as shown below (fig. 68). A and B are the points of emergence of the optic axes, C the point of emergence of the wave-normal making angles of ϕ and ϕ' with OA and OB respectively, O being the origin. Then

$$\phi = r/d \quad \text{and} \quad \phi' = r'/d.$$

The curve on which C must lie therefore satisfies the relation

$$rr' = \text{constant}.$$

This curve is called one of the 'Ovals of Cassini'† and may be found by a simple geometrical construction. If a circle be drawn of radius equal to $\sqrt{DB^2 + rr'}$ with D, the midpoint of AB, as its centre, then every chord such as EBF will give two lines EB, BF such that their product is equal to rr'. If a point C be located so that $AC = EB$ and $BC = BF$, then C must be on the required curve. If this process be repeated for a number of chords

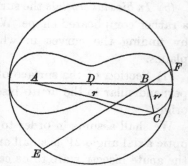

Fig. 68. Construction for the geometrical derivation of the form of certain biaxial optic pictures.

such as EF we obtain the complete oval. The three curves

† H. Lamb, *Infinitesimal Calculus*, p. 320 (Cambridge Univ. Press, 1921).

in fig. 69 correspond to retardations of λ, 2λ and 3λ respectively and have been obtained by the above graphical construction. It will be noticed how closely the figure corresponds to a typical biaxial optic picture.

(2) Section of the surfaces of equal retardation by a plane perpendicular to an optic axis but not passing through the origin.

Very close to one optic axis, say B (fig. 68), the angle ϕ, and therefore r, is practically constant. Since for a surface of

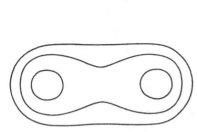

Fig. 69. The form of the biaxial optic picture derived by the construction.

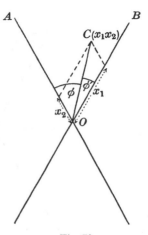

Fig. 70

equal retardation rr' is fixed, r' must also be constant. Near one optic axis the intersection of each surface of equal retardation with this plane is therefore a circle.

(3) Section of the surfaces of equal retardation by the optic axial plane.

Suppose OA, OB in fig. 70 are the directions of the optic axes and C a point on a surface of given retardation. Then the retardation is given by

$$n\lambda = OC(\mu' - \mu'') = OC(\alpha_3 - \alpha_1)\sin\phi \sin\phi',$$

where ϕ and ϕ' are the angles OC makes with OB and OA

respectively. If x_1, x_2 are the coordinates of C referred to OA and OB as axes, then

$$\frac{x_2}{\sin \phi'} = \frac{OC}{\sin 2V},$$

$$x_2 = \frac{OC}{\sin 2V} \sin \phi';$$

similarly,

$$x_1 = \frac{OC}{\sin 2V} \sin \phi.$$

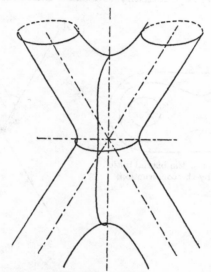

Fig. 71. The general form of the surface of equal retardation for a biaxial crystal. (Taken from *Lehrbuch der Kristalloptik*, F. Pockels (Teubner, Leipzig and Berlin, 1906), with kind permission of the publisher.)

Hence

$$x_1 x_2 = \frac{OC^2}{\sin^2 2V} \sin \phi' \sin \phi$$

$$= \frac{n\lambda \cdot OC}{(\alpha_3 - \alpha_1) \sin^2 2V}.$$

Thus on the parts of the curve where the length of OC is not changing much, and, therefore, $x_1 x_2$ is constant, the form of the curve is hyperbolic having the optic axes as asymptotes. This approximation will always hold in directions inclined at small angles to the acute and obtuse bisectrices and for large values of OC.

A perspective drawing of a typical surface of equal retardation is shown in fig. 71 and from it may be seen the general characters of the sections discussed in paragraphs (1), (2) and (3).

2.3. Determination of the sign of a uniaxial crystal by observations in convergent light.

A common method of determining the optic sign of doubly refracting crystals utilises the change in appearance of an optic picture when a quarter-wave plate of mica or gypsum is superimposed on the crystal plate. The optic picture of a uniaxial crystal cut perpendicular to the optic axis consists of brushes and rings. When a quarter-wave plate with its vibration directions at 45° to the vibration directions of the nicols is placed above the crystal plate, the rings break up into four segments; one pair of opposite segments appears to move inwards, the other pair outwards, by a distance corresponding to a change of path difference of $\lambda/4$. This is illustrated in fig. 72. The explanation of this phenomenon given in several text-books is incomplete, and fails to account for the fact that the dark quadrants are still portions of circles.

Certain rather remarkable properties of this combination of crystal section and quarter-wave plate are seldom mentioned. For instance, the appearance of the picture remains unchanged if the analyser and mica plate are rotated together, provided their vibration directions are always at 45° to each other. Similarly, if the quarter-wave plate is placed *under* the crystal section the phenomenon is the same, and is unchanged when the polariser is rotated with the quarter-wave plate.† These phenomena must clearly be associated with circularly polarised light, otherwise rotating the analyser and

---

† When testing this statement it is necessary to use a plate which has been accurately cut to retard a quarter of a wave-length.

quarter-wave plate together would introduce some alteration in the appearance of the picture.

An analytical treatment of the passage of circularly polarised light through a crystal plate is given in several text-books.† The analysis shows that the intensity of light emerging from the analyser is a fraction x of that entering the crystal, where

$$x = 1 - \sin 2\alpha \sin \delta,$$

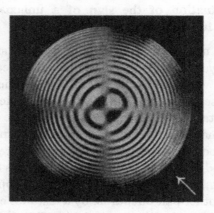

Fig. 72. Photograph of a uniaxial optic picture obtained by superimposing a $\lambda/4$ plate on the section. The slow vibration direction of the $\lambda/4$ plate is parallel to the arrow. Note that the intensity of the light round any circle concentric with the centre of the picture varies continuously, if at all, and that there are no abrupt changes of intensity where the rings pass over the grey cross. (Reproduced from *A Manual of Petrographic Methods* by A. Johannsen (McGraw-Hill, New York and London, 1918), by kind permission of the publishers.)

α being the angle between one vibration direction of the crystal and the vibration direction of the light transmitted by the analyser. δ is the phase difference due to the double refraction introduced in passing through the crystal. In convergent light δ is the same for all rays forming a given circle in the optic picture. If, therefore, we choose a circle for which $\sin \delta = 1$ (i.e. one corresponding to a path difference of $n\lambda + \lambda/4$, where

† T. Preston, *The Theory of Light*, p. 413, 4th edition (Macmillan, London, 1912); A. Schuster and J. W. Nicholson, *An Introduction to the Theory of Optics*, p. 217 (Arnold, London, 1928).

n is an integer), x goes through all values from 0 to 2 as α changes from $+45°$ to $-45°$. For all points of the optic picture lying on a line through the centre and parallel to the plane of vibration transmitted by the analyser $\alpha = 0°$ and hence $x = 1$. Similarly, if $\alpha = 90°$, $x = 1$; hence it follows that the intensity round a circle, concentric with the centre of the picture, waxes and wanes, being a maximum or a minimum along the 45° directions. For $\delta = 0°$ or $180°$, $x = 1$ for all values of α, and there is no change in intensity round the corresponding ring.

We are now in a position to understand the true nature of the optic picture when a $\lambda/4$ plate is superimposed on the crystal plate. The rings are truly circular but do not end abruptly as so many drawings suggest; instead they fade gradually into the grey cross which replaces the black one. Another common error in the representation of this optic picture is to draw the rings with uniformly increasing radii. Like the rings in the ordinary uniaxial optic picture or in Newton's rings formed between a lens and a plate, the rings in monochromatic light have radii approximately proportional to the square root of the number of the ring, counting outwards from the centre.

The sign of the crystal shown in fig. 72 is negative (the arrow indicates the vibration direction of the slow ray of the mica plate). The radial vibrations must be fast, because in the two quadrants where the radial vibration is parallel to the slow vibration of the mica plate, the double refraction of the plate is apparently reduced. The radial vibration of a uniaxial optic picture is carried by the extraordinary ray, and if this is faster than the ordinary ray the crystal has a negative sign. A convenient way of memorising this is to notice that the line joining the innermost black spots makes a negative sign relative to the slow vibration direction of the quarter-wave plate.

3. Internal and external conical refraction.

The terms 'internal' and 'external conical refraction' which are applied to certain optical phenomena are misnomers. The phenomena of internal and external conical refraction, usually demonstrated with plates of aragonite, are due to ordinary double refraction, and the 'explanations' given in nearly all text-books dealing with the subject are incorrect.

3.1. Experimental observations.

The phenomena may be observed by mounting a metal foil (A) pierced by very fine holes in front of a block of aragonite. Its orientation is given in fig. 73. The aragonite is rotated about an axis parallel to the third mean line. An adjustable eyepiece E is focussed to give a clear image of the pinholes seen through the crystal. In general two images of each pinhole are observed.

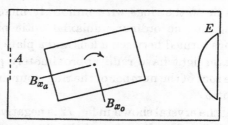

Fig. 73. Showing the arrangement with a perforated plate and an aragonite block for observing double refraction in directions nearly parallel to an optic axis.

But if the light travels through the crystal in a direction nearly parallel to an optic axis then the spots grow into crescent-shaped figures. As the light beams by which the pinholes are seen become less inclined to the optic axis, the 'wings' grow, until the two images of any one pinhole coalesce to give a complete luminous circle. This experiment is usually but incorrectly regarded as a demonstration of internal conical refraction.

Lloyd's experiment was supposed to demonstrate external conical refraction. For this purpose two pieces of tinfoil each pierced with a fine hole were placed one on each side of a block of aragonite so that the line joining the holes was practically parallel to an optic biradial. A conical beam of light was allowed to fall on the crystal at the appropriate angle so that the light was refracted along the optic biradial. The light emerging from the second pinhole was distributed uniformly round the surface of a cone, as could be seen by receiving it on a screen. The diameter of the circle was proportional to the distance of the screen from the nearer piece of tinfoil. This phenomenon, like that described above, depends simply on double refraction and not on external conical refraction.

The fact is often overlooked that a single ray of light or even a strictly parallel beam is unobtainable. It is easy to convince oneself of this by imagining an attempt to make a beam of light emerging from a collimator more and more parallel by gradually closing the entrance aperture. As the Irishman said, only when the aperture was completely shut would the light be strictly parallel.† Consequently, explanations of internal and external conical refraction which postulate wave-normals travelling exactly along the optic axis or rays directed exactly along the optic biradial are false, because if such wave-normals and rays existed they would be invisible.

3.2. Explanation of the effects in terms of double refraction.

We shall employ Sylvester's construction to explain what is observed when conical beams of small angle pass through the crystal in directions nearly parallel to an optic axis. The diagram in fig. 74 is a stereographic projection using a projection sphere of large radius, because the angular distances involved in this study are all very small, of about 1° or less. Suppose the optic axis emerges at P and the optic biradial at Q, and that the acute bisectrix emerges a long way to the left of the paper on the line PQ produced. Consider a wave-normal R close to P. To find the rays R_1, R_2 associated with R, join R to P and to the other optic axis P'. P' is so far from P

† For the purpose of this discussion we can neglect diffraction effects.

that in this diagram the line RP' is parallel to PQ. The lines RR_1, RR_2 bisect the angle PRP' internally and externally. According to Sylvester's construction QS is drawn perpendicular to RR_1 so that $QR = RS$, and then S is joined to the second optic biradial Q'. Q' is so far off that SR_1 is taken parallel to PQ. R_1 then gives the direction of one ray associated with the wave-normal R.

It is easy to show by a little geometry that QR_1 is parallel to RP, and that

$$QR_1 = QU = QP - UP = QP - PR.$$

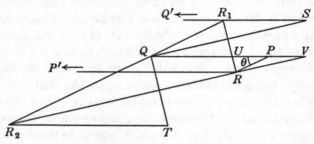

Fig. 74. Diagram showing the relation between the directions of the wave-normal R and associated rays R_1, R_2 when R is nearly parallel to an optic axis, P.

The second ray R_2 associated with the wave-normal R may be found in the same way as indicated in fig. 74. Again, it follows from the geometry of the figure that QR_2 is parallel to RP and that

$$QR_2 = QV = QP + PV = QP + PR.$$

Also R_2, Q and R_1 are collinear.

So far we have considered only abstractions, such as isolated rays and wave-normals. Now let us consider an actual beam of divergent light uniformly distributed within a small-angled cone MRN. The rays associated with this cone of wave-normals are indicated in fig. 75 by the shaded areas, M_1N_1, M_2N_2. Since for any arbitrary wave-normal R, for which the corresponding rays are R_1 and R_2,

$$QR_2 = QP + PR,$$

the rays in the region M_2N_2 lie at distances from Q differing only by the diameter of the cone R. But since every line such as PR is parallel to the corresponding line QR_2, the rays within M_2N_2 are distributed over the elongated area between QM_2 and QN_2, where QM_2 and QN_2 are parallel to PM and PN respectively. The nearer the axis of the small cone approaches the optic axis P the greater the arc over which the areas M_1N_1, M_2N_2 extend, until, in the limit, when the axis of the

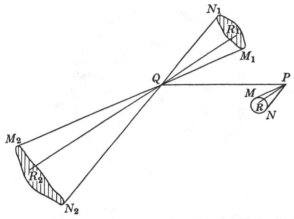

Fig. 75. Diagram showing the directions of ray bundles M_1N_1, M_2N_2 which arise from a cone of wave-normals MN when the latter are directed nearly parallel to the optic axis, P.

cone coincides with OP, the rays lie in uniform circles concentric with the optic biradial Q. If the crystal plate is plane parallel, the rays of light which form the circles travel parallel to their original direction of propagation when they emerge from the crystal. Thus the small-angled cone which gives rise to the circles of rays is changed into an approximately cylindrical beam of light the inner portion of which is convergent and the outer divergent. The total divergence of the light is, from what has already been said, the same as on entry into the crystal. The variation of the intensity of the light along any radius of the rings may be obtained as follows. The wave-normals within the annulus about P (fig. 76) of radii r and $r + dr$

give rise to rings of light of radii $QP+r$, $QP+r+dr$ and $QP-r$, $QP-r-dr$, concentric with Q. The amount of light in the ring about P is $2\pi r\, dr\,.\,I$, where I is proportional to the light intensity per unit solid angle of the cone about P. The light intensity in each of these rings concentric with Q, if we neglect the difference in their radii, is

$$\frac{\pi r\, dr\,.\,I}{2\pi\,.\,PQ\,.\,dr} = \frac{r}{2PQ}\,I.$$

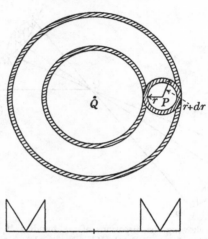

Fig. 76. Diagram illustrating the determination of the radial distribution of the light in the luminous rings produced by double refraction near an optic axis.

Thus the intensity of the light on a circle of radius $PQ+r$ decreases to zero as r falls to zero, and if a curve be drawn showing the intensity distribution across the luminous circle it would be ideally as indicated in fig. 76. In practice it is impossible to realise exactly the conditions we have assumed, but it is quite possible to observe the dark line round the centre of the luminous circle. The dark circle corresponds to light which has suffered internal conical refraction, and we see that such light is characterised by its invisibility. This is what we expect from the statement made earlier in this section, viz. that a strictly parallel beam of light is invisible.

By the same analysis it may be shown that a cone of rays travelling near one of the optic biradials gives wave-normals which lie within circles concentric with the optic axis. In the classical experiment of Lloyd an emergent beam which was practically parallel was received on a reflecting screen. To obtain such an effect it is necessary to limit appropriately the divergence of the incident beam.

3.3. The angle of the cones of internal and external conical refraction.

If it is assumed, as it is here, that the angle between each optic axis and its neighbouring optic biradial is small compared with the optic axial angle, $2V$, then except for the effect of refraction at the crystal surface, the angles of the cones of internal and external conical refraction are equal. The error in this statement is small and a more exact analysis may be found in Liebisch.† The angle 2∂ of either cone is equal to twice the angle between the optic axis and optic biradial, and 2∂ may be shown to be

$$\frac{2\sqrt{(\alpha_2 - \alpha_1)(\alpha_3 - \alpha_2)}}{\alpha_2}.$$

The following table gives the magnitude of 2∂ for a few minerals.

TABLE XXIV. *Angle of the cone of internal conical refraction*

Mineral	2∂
Anhydrite	1° 5′
Aragonite	1° 47′
Chrysoberyl	0° 13′
Diopside	0° 52′
Mica	0° 59′
Orthoclase	0° 11′
Sulphur	7° 22′

† Th. Liebisch, *Physikalische Krystallographie*, p. 342 (von Veit, Leipzig, 1891).

3.4. Polarisation of the rays and waves in conical refraction.

It has been shown to be a property of the indicatrix that each ray has a vibration direction lying in the plane containing the ray itself and the associated wave-normal. If we consider the rays emerging at points round the luminous circle corresponding to the cone of internal conical refraction, then we see that the vibration directions must be arranged as shown in fig. 77, where P represents the optic axis and Q the optic biradial. Similarly, the vibration directions associated with light travelling in directions close to the cone of external conical refraction are as shown in fig. 78.

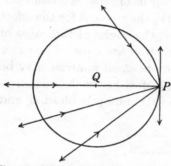 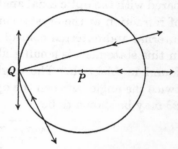

Fig. 77. Orientation of the vibration directions relative to the optic axis P and the optic biradial Q round the cone of rays corresponding to a wave-normal directed along P.

Fig. 78. Orientation of the vibration directions relative to P and Q of the wave-normals corresponding to a ray directed along Q.

4. Rotatory polarisation.

4.1. Experimental work on the rotatory polarisation of biaxial crystals.

Investigations on the optical activity of biaxial crystals have not extended beyond determining the angle of rotation of the plane of polarisation when light travels parallel to an optic axis. The technique used is just the same as for the investigation of uniaxial optically active crystals, namely, a plane parallel plate cut perpendicular to an optic axis is viewed in convergent polarised light between crossed nicols. The optic picture for optically active biaxial crystals differs from that for inactive crystals in that the brush fades away at the centre of each eye instead of going right through it.† If a quarter-wave plate be inserted at the 45° position with respect to the vibration direction of the analyser so that only circularly polarised light is transmitted to the eye, then the brush disappears and the rings form a single continuous spiral† wound in a direction depending on the sense of rotation of the transmitted circularly polarised light. The results obtained are classified below according to the crystal class of the substance.

Class 2.

Substance	Rotation per mm. (Na-D light)‡		Orientation of optic axial plane
	Optic axis 1	Optic axis 2	
Cane sugar,§ $C_{12}H_{22}O_{11}$	− 1·6°	+ 5·4°	‖(010)
Rhamnose,† $C_6H_{12}O_5.H_2O$	+12·9°	+ 5·4°	‖(010)
Quercitol,¶ $C_6H_7(OH)_5$	− 3·5°	− 4·6°	‖(010)
Tartaric acid,†† $C_4H_6O_6$	+10·8°	+10·8°	‖[010]

It will be noticed that where the symmetry allows it the rotations per mm. about the axes are different both in amount, and, in one case, in sign. In tartaric acid the sense and amount of rotation for light travelling along the optic axes are the same, since the latter are related by a diad axis.

Class m.

One substance in this class, namely, mesityl oxide oxalic methyl ester, is known to show rotatory polarisation.‡‡

Class 222.

Substance	Rotation per mm. (Na-D light)
Rochelle salt,† $NaKC_4H_4O_6.4H_2O$	− 1·4°
$NaNH_4C_4H_4O_6.4H_2O$†	+ 1·6°
Epsom salts,§ $MgSO_4.7H_2O$	+ 2·0°
$NaH_2PO_4.2H_2O$†	− 4·4°

† For very clear photographs see M. H. Dufet, *Bull. Soc. Min. France*, **27**, 156 (1904).

‡ Confusion exists in the literature as to the meaning of the sign of optical rotation. In this book a positive or right-handed crystal rotates the plane of polarisation in a direction clockwise relative to an observer looking towards the light source.   § L. Longchambon, *C.R.* **172**, 1187 (1921).

¶ V. V. Karandeev, *Bull. Acad. Sci. Petrograd*, **9**, 1285 (1915).

†† L. Longchambon, *C.R.* **178**, 951 (1924).

‡‡ *Handbuch der Physik*, vol. xx, p. 635 (Springer, Berlin, 1928).

The optic axes are necessarily symmetrical with respect to axes of symmetry, and hence the rotation about each must be the same.

4.2. The use of the phenomenon of rotatory polarisation as a means of deciding the class of a crystal.

With the exception of class m, in which there is one example of an optically active crystal, rotatory polarisation has only been observed in the following enantiomorphous classes. Examples of crystals belonging to these, and the values of the rotation per mm., are given.

		Na light
23–T	Sodium chlorate, $NaClO_3$	3° 8′
42–D_4	Ethylene diamine sulphate, $C_2H_4(NH_2)_2 . H_2SO_4$	15° 30′
6–C_6	Potassium lithium sulphate, $KLiSO_4$	3° 26′
32–D_3	Quartz, SiO_2	21° 44′
	Cinnabar, HgS	325° (red light)
3–C_3	Sodium periodate, $NaIO_4 . 3H_2O$	23° 8′
222–D_2	Epsom salts, $MgSO_4 . 7H_2O$	$+2.0°$
2–C_2	Cane sugar, $C_{12}H_{22}O_{11}$	$-1.6°$ and $+5.4°$

Rotatory polarisation has not been observed in classes $\bar{4}$ and $\bar{4}2m$, though they lack centres of symmetry. On the basis of previous experience we should therefore expect any crystal showing optical activity to belong to one of the enantiomorphous classes. The paucity of the experimental material, and the existence of rotatory polarisation in one non-enantiomorphous class, make it necessary to exercise caution in drawing conclusions as to the symmetry of a crystal from the occurrence of rotatory polarisation.

4.3. The theory of the occurrence of rotatory polarisation in various crystal classes.

It is an experimental fact that the rotation of the plane of polarised light when travelling along a given line is independent of the direction of travel. The simplest general assumption consistent with this fact that we can make about the variation of ρ, the rotation per unit thickness of crystal, is that it depends on the second order terms of the direction-cosines, i.e.

$$\rho = c_{13}^2 r_{11} + c_{23}^2 r_{22} + c_{33}^2 r_{33} + 2c_{13}c_{23}r_{12} + 2c_{23}c_{33}r_{23} + 2c_{33}c_{13}r_{31}.$$
$$\ldots\ldots(1)$$

We may apply this expression to the various classes of symmetry and find the form of the surfaces which represent the variation of ρ with direction.

If the crystal possesses a centre of symmetry, then on interchanging $-c_{13}$, $-c_{23}$, $-c_{33}$ for c_{13}, c_{23}, c_{33} a right-handed rotation must invert into a left-handed one, i.e. ρ must change sign. Since, however, all the terms on the right-hand side of the expression are second order in the c_{ik}'s this is impossible unless $\rho = 0$. Hence we conclude that no crystal possessing a centre of symmetry can show rotatory polarisation. There are therefore twenty-one classes of symmetry which remain to be considered.

Class 1—C_1.

To this class of the anorthic system the general expression (1) applies, and the form of the surface describing ρ (called a gyration surface) depends on the values of the constants r_{ik}.

Class m—C_s.

We shall suppose the plane of symmetry to be perpendicular to the axis X_3. The operation of a plane of symmetry is to reflect a right-handed rotation into a left-handed one. Thus if ρ is positive for a direction defined by c_{13}, c_{23}, c_{33} it must be negative and of equal amount for a direction defined by c_{13}, c_{23}, $-c_{33}$. The expression (1) for ρ must therefore reduce to

$$\rho = 2c_{33}(c_{23}r_{23} + c_{13}r_{31})$$

in order to satisfy this condition.

Class 2—C_2.

Whereas a centre or plane of symmetry converts a right-handed vibration into a left-handed one and *vice versa*, any axis of symmetry leaves the sign of the rotation unchanged.

Hence if the diad axis be made to coincide with X_3 we must be able to exchange $-c_{13}$, $-c_{23}$ for c_{13} and c_{23} without affecting the sign of ρ. The full expression, therefore, reduces to

$$\rho = c_{13}^2 r_{11} + c_{23}^2 r_{22} + c_{33}^2 r_{33} + 2c_{13}c_{23}r_{12}.$$

Class 222—D_2.

Applying the same principle to all three diad axes, we obtain

$$\rho = c_{13}^2 r_{11} + c_{23}^2 r_{22} + c_{33}^2 r_{33}.$$

Class mm—C_{2v}.

We shall suppose the two planes of symmetry to intersect in the X_3 axis. If $-c_{13}$ is exchanged for c_{13} (corresponding to the plane of symmetry perpendicular to the X_1 axis), c_{23} and c_{33} remaining unchanged, the sign of ρ must change. Also when $-c_{23}$ is exchanged for c_{23} (corresponding to the plane perpendicular to X_2), c_{13} and c_{33} remaining unchanged, ρ must change sign. The expression for ρ becomes

$$\rho = 2c_{13}c_{23}r_{12}.$$

Classes 3, 32, 4, 42, 6, 62—C_3, D_3, C_4, D_4, C_6, D_6.

It has been proved in paragraph 7, p. 13, that a second order tensor is represented by a surface of revolution in these classes. No special proof need be given here. Further, the axis of rotation of the surface must coincide with the principal axis of the crystal. Thus $r_{11} = r_{22}$ and $r_{12} = r_{23} = r_{31} = 0$, and hence

$$\rho = (c_{13}^2 + c_{23}^2)\, r_{11} + c_{33}^2 r_{33}.$$

Classes 3m, 4mm, 6mm—C_{3v}, C_{4v}, C_{6v}.

If we suppose a plane of symmetry to be perpendicular to the X_1 axis, then on exchanging $-c_{13}$ for c_{13} the sign of ρ must change. But since the principal axes in classes 3m, 4mm and 6mm are the same as in classes 3, 4 and 6 the required expression for ρ must be some modification of the last equation. As c_{13} only appears as c_{13}^2 in this expression ρ will not change sign on exchanging $-c_{13}$ for c_{13}, and hence

$$\rho = 0.$$

Class $\overline{4}$—S_4.

Directions related by an alternating tetrad axis are

$$(1)\ c_{13},\, c_{23},\, c_{33}; \qquad (2)\ c_{23},\, -c_{13},\, -c_{33};$$
$$(3)\ -c_{13},\, -c_{23},\, c_{33}; \qquad (4)\ -c_{23},\, c_{13},\, -c_{33}.$$

On exchanging (1) for (2) ρ must change sign, because the change is equivalent to a rotation of 90° about the tetrad axis together with a reflection in a horizontal plane of symmetry. Exchanging (1) for (3) is equivalent to a diad axis parallel to X_3, and the operation of this symmetry element will leave the sign of ρ unchanged. Thus ρ is given by

$$\rho = (c_{13}^2 - c_{23}^2)\, r_{11} + 2c_{13}c_{23}r_{12}.$$

Class $\overline{4}2m$—D_{2d}.

A further condition, in addition to those required by the symmetry of class $\overline{4}$, is that ρ shall not change sign on inserting c_{13}, $-c_{23}$, $-c_{33}$ in place of c_{13}, c_{23}, c_{33} (corresponding to a diad axis in the X_1 direction). Hence

$$\rho = (c_{13}^2 - c_{23}^2)\, r_{11}.$$

Classes $\overline{6}$, $\overline{6}2m$—C_{3h}, D_{3h}.

Both of these classes contain as symmetry elements a trigonal axis perpendicular to a horizontal plane of symmetry. The trigonal axis necessitates that the surface representing ρ shall be one of revolution about X_3, but the horizontal plane of symmetry requires that on exchanging c_{13}, c_{23}, $-c_{33}$ for c_{13}, c_{23}, c_{33}, ρ shall change sign. Now ρ in class 3 is given by

$$\rho = (c_{13}^2 + c_{23}^2)\, r_{11} + c_{33}^2 r_{33},$$

and it is clear that the above exchange does not change the sign of ρ, hence $\rho = 0$.

Classes 43, 23—O, T.

If in the expression of ρ in class 222 we insert the condition imposed by the trigonal axes of a cubic crystal, namely, that an equivalent state is obtained when the three axes are cyclically interchanged, then

$$\rho = r_{11}(c_{13}^2 + c_{23}^2 + c_{33}^2) = r_{11}$$

and the specific rotation is the same in all directions.

Class $\overline{4}3m$—T_d.

This class is obtained from class 23 by the addition of a plane of symmetry. Thus, for directions equally inclined to planes of symmetry ρ must change sign. This is incompatible with the previous result, namely, that ρ is independent of direction. Hence $\rho = 0$. The information concerning the variation of the expression for ρ with the symmetry of the crystal is summarised below.

TABLE XXV. *The variation of the rotatory polarisation with direction in the classes lacking a centre of symmetry*

Crystal system	Class	Expression for ρ
Triclinic	1	$c^2_{13}r_{11} + c^2_{23}r_{22} + c^2_{33}r_{33} + 2c_{13}c_{23}r_{12} + 2c_{23}c_{33}r_{23} + 2c_{33}c_{13}r_{31}$
Monoclinic	m	$2c_{33}(c_{23}r_{23} + c_{13}r_{31})$
	2	$c^2_{13}r_{11} + c^2_{23}r_{22} + c^2_{33}r_{33} + 2c_{13}c_{23}r_{12}$
Rhombic	222	$c^2_{13}r_{11} + c^2_{23}r_{22} + c^2_{33}r_{33}$
	mm	$2c_{13}c_{23}r_{12}$
Rhombohedral	3	$(c^2_{13} + c^2_{23})r_{11} + c^2_{33}r_{33}$
	32	$(c^2_{13} + c^2_{23})r_{11} + c^2_{33}r_{33}$
	3m	0
Tetragonal	4	$(c^2_{13} + c^2_{23})r_{11} + c^2_{33}r_{33}$
	42	$(c^2_{13} + c^2_{23})r_{11} + c^2_{33}r_{33}$
	4mm	0
	$\overline{4}$	$(c^2_{13} - c^2_{23})r_{11} + 2c_{13}c_{23}r_{12}$
	$\overline{4}2m$	$(c^2_{13} - c^2_{23})r_{11}$
Hexagonal	6	$(c^2_{13} + c^2_{23})r_{11} + c^2_{33}r_{33}$
	62	$(c^2_{13} + c^2_{23})r_{11} + c^2_{33}r_{33}$
	6mm	0
	$\overline{6}$	0
	$\overline{6}2m$	0
Cubic	43	r_{11}
	23	r_{11}
	$\overline{4}3m$	0

4.4.* The transmission of light through quartz in directions inclined to the optic axis.†

If a beam of plane polarised light falls normally on a plate of a uniaxial optically inactive crystal cut with a general orientation relative to the optic axis, the light emerging from it will, in general, be elliptically polarised. Two positions may, however, be found on rotating the crystal about the light beam as axis, in which plane polarised light is transmitted unchanged. With a quartz plate of general orientation it is not possible as a rule to find any such positions. Only elliptically polarised light is transmitted unchanged, except when the plate is cut with a particular orientation to be discussed later. The elliptical vibrations have their major axes either parallel or perpendicular to the principal plane, and the ellipticity must have a definite value depending on the inclination of the wave-normal to the optic axis. The two ellipses defining this state of polarisation are similar to one another but have opposite senses of rotation and their major axes coincide with the vibration directions which the crystal would have if it were not active. In a left-handed crystal the ellipse with the right-handed sense of rotation‡ corresponds to the faster ray. Since quartz is optically positive this is the ordinary ray and the major axis of the ellipse is perpendicular to the plane containing the ray and the optic axis. The ellipse with the right-handed sense of rotation has its major axis lying in this plane. The ellipticity χ (ratio of minor to major axis) changes rapidly with the inclination ϕ of the light ray to the optic axis, and depends only on this angle, as shown in Table XXVI.

It will be seen that the ellipticity decreases rapidly near to the optic axis, and becomes zero when $\phi = 56° \ 10'$ and then

† G. Bruhat and P. Grivet, *J. Phys. Radium*, **6**, 12 (1935); C. Munster and G. Szivessy, *Phys. Z.* **36**, 101 (1935).

‡ The ellipse is supposed located at the surface of the crystal where the light emerges and its sense of rotation is that seen by an observer looking towards the light source.

changes sign and increases up to 90°. Therefore, *only in directions inclined at 56° 10′ to the optic axis is a quartz crystal capable of transmitting a plane polarised beam unchanged.* This is of importance in the design of very precise optical instruments. The ratio, $\rho_\perp/\rho_\parallel$, of the two rotatory powers does not vary with

TABLE XXVI. *Variation of the ellipticity of the polarised light which is transmitted unchanged with the inclination of the light to the optic axis*†

ϕ	χ
0°	1·0
12° 2′	0·0987
18° 22′	0·0391
24° 20′	0·0198
32° 33′	0·00911
46° 30′	0·00218
56° 10′	0·00000
90° 0′	−0·00203

wave-length between 5461 and 2653 A.U. within the experimental error, and is equal to −0·53. It is difficult to measure the rotatory power of quartz in directions perpendicular to the optic axis because the measurement involves the determination of the ratio of the minor to the major axis of elliptically polarised light when this ratio is only about 1/500. This problem is even more complicated with biaxial crystals, and this probably accounts for the fact that observations on them have so far been restricted to the directions of the optic axes where the double refraction is unimportant.

5. Optical properties of metallic and semi-metallic crystals.

5.1. Introduction.

The theoretical study of the reflection of light from opaque substances was practically completed at the beginning of the century. During the last fifteen years metallic and semi-metallic ores have been identified in polished sections by a determination of the intensity of the light reflected. In this

† G. Szivessy and C. Münster, *Ann. Physik*, **20**, 703 (1934).

field the study has its greatest technical importance,† but its scientific interest is much wider.

The term 'refractive index' has a rather different meaning when applied to opaque substances than when applied to transparent substances. The relation between the angle of incidence and the angle of refraction,

$$\frac{\sin i}{\sin r} = \mu,$$

which defines the refractive index of a transparent substance, cannot be directly applied to an opaque one, since no angle of refraction can be measured. In the analysis of the reflection and refraction of light in opaque substances, two constants appear, viz. the refractive index, n, and the absorption coefficient, κ. n defines the relation between the angle of incidence and the imaginary angle of refraction. The absorption coefficient determines the depth to which the light can penetrate into the opaque substance. When κ becomes indefinitely small the expression for n becomes the same as in transparent substances. The values for n have a wide range, e.g. 5 for stibnite, 0·005 for sodium, and so have the values for κ, e.g. 6 for magnesium, 0·11 for haematite.

5.2. The principal phenomena of the reflection of light from metallic substances.

If a beam of plane polarised light is allowed to fall on an optically flat metallic surface, reflection occurs so that the angle of reflection is equal to the angle of incidence. The reflected light is, in general, not plane but elliptically polarised. The ellipticity of this polarisation depends on the substance reflecting, the angle of incidence, and the polarisation azimuth of the incident light. This last is denoted ψ_E and is equal to the complement of the angle between the plane of

† F. E. Wright, *Proc. Amer. Phil. Soc.* **58**, 401 (1919); F. E. Wright, *Mining and Metallurgy*, No. 158 (1920); M. J. Orcel, *Bull. Soc. Min. France*, **53**, 301 (1930); A. Cissarz, *Z. Krist.* **78**, 445 (1931).

vibration of the incident light and the plane of incidence. The experimental method which has usually been employed for studying these properties of metals is shown in fig. 79. A spectrometer is provided with a polariser, P, analyser, Q, and a Babinet compensator, B. The polarising prisms and compensator are placed in the parallel light beams between the reflecting surface, M, and the objective lens of the collimator, A, and the telescope, T. Light enters through a small circular aperture in the collimator A, is rendered plane polarised by passage through P, is reflected at M, then passes through

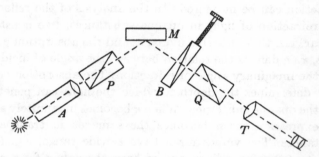

Fig. 79. The experimental arrangement used for investigating the reflection of light from metals: A, collimator; P, polariser; M, reflecting plate; B, Babinet compensator; Q, analyser; T, telescope.

B, Q, and finally the telescope. The nicol prisms P and Q are provided with graduated scales and the screw of the compensator is graduated so that the wedges may be given any desired relative displacement. The usual graduated circle of the spectrometer makes it possible to rotate M or B, Q, and T together about the vertical axis. The light after reflection at M is generally elliptically polarised, but by suitable adjustment of B it can be rendered plane polarised. If it is plane polarised then the light can be extinguished by a suitable setting of Q.

An elliptical vibration, such as that represented in fig. 80, can be resolved into two perpendicular components Oa, Ob with a certain phase difference between them. After adding or subtracting the appropriate phase difference the elliptical vibration is converted into a plane polarised vibration parallel to Oc. The tangent of the angle ψ_R gives the ratio of the components Oa, Ob.

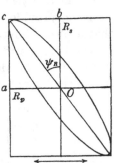

We shall now consider three experiments with a reflecting plate of isotropic metal which enable n and κ to be found.

(1) The nicol P is set to transmit light having a vibration direction parallel to the axis about which M turns ($\psi_E = 0$). The light reflected from M is examined through B, Q, T for various angles of incidence. It will be found that with B set to introduce no phase difference the setting of Q necessary to produce extinction will be the same for all angles of incidence. Thus the reflected light is plane

Plane of incidence

Fig. 80. The relation of elliptically polarised light to the plane polarised light into which it may be converted by a Babinet compensator.

polarised with a vibration direction parallel to the plane of the metal surface.

(2) The polariser is set to transmit no vibrations except those lying in the plane of incidence ($\psi_E = \pi/2$). Again extinction will be obtained in T if B is arranged to introduce no phase difference, showing that for all angles of incidence the light is plane polarised. The difference between this experiment and the last is that the intensity of the reflected light decreases as the angle of incidence increases, until, at a certain angle, $\bar{i}$, known as the principal angle of incidence, a minimum intensity is observed. The curves of fig. 81 give the variation with angle of incidence of the intensity of the light, having vibration directions in and perpendicular to the plane of incidence, for copper and galena, PbS.

(3) The polariser is now set to transmit light vibrating in a plane at 45° to the plane of incidence. At grazing, and at normal, incidence the reflected light is plane polarised. When the light is reflected at grazing incidence the analyser must be

crossed with respect to the polariser to produce extinction. To produce extinction when the angle of incidence is nearly zero the analyser must be rotated through 90° from this position, provided B is left unchanged. This rotation shows that a phase difference of π has been introduced between the perpendicular and parallel components of the incident light. For angles of incidence between 0° and 90° the reflected light is elliptically polarised. The phase difference $\varDelta$ between the

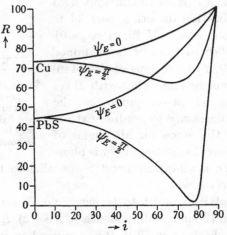

Fig. 81. Showing the variation with angle of incidence of the intensity of light reflected from surfaces of copper and galena for two azimuths of the plane of polarisation of the incident light. (Reproduced from the *Zeitschrift für Kristallographie*, Bd. 76, p. 426 (1931), by kind permission of the publishers.)

components parallel and perpendicular to the plane of incidence may be determined by the displacement of the compensator. The complement of the angle which the vibration direction of the light emerging from B makes with the plane of incidence is denoted ψ_R. The ratio between the amplitudes of the components vibrating parallel (R_p) and perpendicular (R_s) to the plane of incidence is then given by the tangent of the angle ψ_R (see fig. 80). The curve showing the variation with the angle of incidence i of the phase difference $\varDelta$ between the components R_s and R_p is given in fig. 82. At the angle $\bar{i}$, when

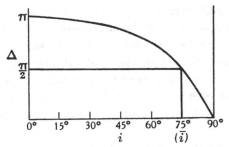

Fig. 82. The variation with angle of incidence of the phase difference of the reflected components.

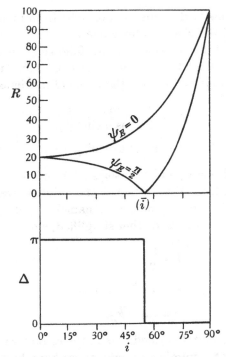

Fig. 83 (upper half). The variation with angle of incidence of the reflecting power of a transparent medium for two azimuths of the planes of polarisation of the incident light.

(Lower half) The corresponding variation of the phase difference between the components of the reflected light polarised in and perpendicular to the plane of incidence. (The upper half of the figure is reproduced from the *Handbuch der Physik*, Bd. 20, p. 211, Fig. 12 (Springer, Berlin, 1928), by kind permission of the publishers.)

$\tan \psi_R$ is a minimum $\varDelta = \pi/2$. It is illuminating to compare the curves of figs. 81, 82 with the corresponding ones for transparent media, which are given in fig. 83. According to Brewster's law the intensity of the light vibrating in the plane of incidence decreases to zero at the polarising angle, and hence the curve for reflecting power descends to zero. Also at this angle there is an abrupt change of phase from π to 0.

5.3. Determination of the refractive index and absorption index.

The theories of the reflection of light from metals are complicated and the rigorous formulae long.[†] For most metallic and semi-metallic substances the following approximations may be used, when the incident light has a vibration direction making an angle of 45° with the plane of incidence:

$$n = \frac{\sin i \tan i \cos 2\psi_R}{1 + \cos \varDelta \sin 2\psi_R},$$

$$\kappa = \sin \varDelta \tan 2\psi_R.$$

If a series of corresponding values of $\varDelta$ and ψ_R are obtained, when i varies from 0° to 90°, then the values corresponding to the principal angle of incidence, namely, $\bar{\imath}$, $\bar{\varDelta}$, $\bar{\psi}_R$, may be found and the formula further simplified, since

$$\varDelta = \frac{\pi}{2},$$

$$n = \sin \bar{\imath} \tan \bar{\imath} \cos 2\bar{\psi}_R,$$

$$\kappa = \tan 2\bar{\psi}_R.$$

Using the method described above, the refractive and absorption indices of a number of substances have been found and are given in the accompanying table.[‡]

[†] *Handbuch der Physik*, Bd. 20, p. 240 (Springer, Berlin, 1928); P. Drude, *Theory of Optics* (Longmans, Green and Co. London, 1922).

[‡] M. Born, *Optik*, p. 265 (Springer, Berlin, 1933).

TABLE XXVII. *The refractive index, absorption index and percentage reflecting power of various metals*

Substance	n	κ	$r \%$
Silver	0·18	20·2	95·0
Magnesium	0·37	11·9	92·9
Gold	0·37	7·6	85·1
Aluminium	1·44	3·6	82·7
Tin	1·48	3·55	82·5
Cadmium	1·13	4·4	84·7
Mercury	1·62	2·72	75·3
Steel	2·27	1·48	58·9
Lead	2·01	1·73	62·1
Bismuth	1·78	1·57	54·3

5.4. Measurement of the intensity of light reflected from metallic surfaces at normal incidence.

The connection between the reflecting power R (the ratio of reflected to incident intensity) and the optical constants is given by

$$R = \frac{(n-1)^2 + n^2\kappa^2}{(n+1)^2 + n^2\kappa^2}.$$

A high reflecting power has no direct relation to either n or κ, since it may be due to the combination of either a high absorption index with a low refractive index, or a low absorption index with a high refractive index. The following examples illustrate this. The reflecting power of silver is 95 per cent, its refractive index 0·18 and its absorption index 20·3, while platinum with a reflecting power of 70 per cent has refractive and absorption indices equal to 2·06 and 2·1, respectively. In the expression for R, n is the refractive index of the metal relative to the isotropic medium bounding it. Now if this medium were replaced by a liquid the reflecting power would be changed. In principle it is possible to make use of this fact to find n and κ, but the accuracy with which R can be measured is insufficient to permit this. The reflecting power is influenced by the method of polishing and by films of grease or other material which may contaminate the surface.

The production of a sufficiently large homogeneous surface for examination in the way described above is usually difficult. The crystals to be examined are often small, intimately twinned, and intergrown with others. Hence the study of metallic ores is most conveniently carried out under the microscope.

Apparatus.

The determination of R can be made in a variety of ways. A method which is much favoured by petrologists is due to Berek.† A microscope is furnished with a device for splitting the incoming light into two beams, one of which is reflected from the crystal while the other traverses two nicols, placed one behind the other. The first nicol is fixed, the second can rotate. The second transmits a variable part of the light passed by the first, the fraction being measured by the setting of the second nicol relative to the first. The two halves of the original beam are compared in the field of the eyepiece and brought to equality by turning the second nicol. In this way the fraction of light reflected by the crystal is found. Other methods employ photoelectric or selenium cells to compare the two beams of light. But whatever the method used for measuring the intensity the microscope must be provided with a polariser through which the light must pass before being reflected from the crystal, and an analyser for studying the reflected light.

5.5. Relation between the symmetry and the reflection of light at normal incidence from crystal sections.

Whereas the characteristics of the transmission of light through transparent media can be represented by the optical indicatrix, any section of which has two planes of symmetry, the surfaces representing refractive and absorption indices and reflecting power are of higher order than the second. These sections usually have no planes of symmetry.

† H. Schneiderhöhn and P. Ramdohr, *Lehrbuch der Erzmikroskopie*, Bd. 1, p. 157 (Borntraeger, Berlin, 1934).

Isotropic opaque substances have the same reflecting power for all orientations of the crystal relative to the plane of polarisation; between crossed nicols there is either extinction, or, should the value of the absorption index be high, uniform illumination in all orientations.

Basal sections of uniaxial crystals behave as though they were isotropic between crossed nicols; for biaxial crystals there are four sections from which circularly polarised light is reflected unchanged†; other sections give elliptically polarised reflections. The complete theory of the variation of the refractive and absorption indices with vibration direction in birefringent crystals is complicated, and will not be considered here.

Unlike transparent crystals, the maximum amount of light need not pass the analyser when the crystal is in the 45° position, and the colours in the two perpendicular 45° positions are often different.

5.6. Glossary of terms.

Plane of incidence: plane containing light ray and normal to the reflecting surface.

Angle of incidence: angle between the light ray and normal to reflecting plane.

Plane of vibration: (1) if the light is extinguished by using a nicol prism the plane of vibration of the incident light contains the direction of propagation of the light and the long diagonal of the nicol; (2) in theoretical discussion the plane of vibration contains the electric vector and the direction of propagation of the light.

Refractive index, n: a constant appearing in the equations derived from the theory of the reflection of light from opaque substances which becomes equal to $\sin i/\sin r$ when the absorption coefficient is infinitely small.

† H. Schneiderhöhn and P. Ramdohr, *Lehrbuch der Erzmikroskopie*, Bd. 1, p. 133 (Borntraeger, Berlin, 1934).

Absorption index, κ: a constant which determines the absorption of light in the reflecting substance.

ψ_E = the complement of the angle between the plane of incidence and the plane of vibration of the incident light (i.e. the plane containing the short diagonal of the polarising nicol).

Δ = phase difference in the reflected light between the two components $\parallel$ and $\perp$ to the plane of incidence.

ψ_R = complement of the angle between plane of vibration of the light which has been rendered plane polarised by the compensator, and the plane of incidence.

$\bar{i}$ = principal angle of incidence, i.e. the angle of incidence at which the ratio between the amplitude of the reflected-light vibrations in and perpendicular to the plane of incidence is a minimum.

R_s = amplitude of light vibrating perpendicular to the plane of incidence.

R_p = amplitude of light vibrating parallel to the plane of incidence.

6.1. The relation between the refractive indices of crystals and the nature of the constituent atoms or ions.

It has long been recognised that, in general, crystals containing heavy elements have high refractive indices. The relation is most clearly demonstrated in isomorphous series in which some ion is successively replaced by heavier elements. For example, in the series of the alkaline earth carbonates cerussite, $PbCO_3$, and witherite, $BaCO_3$, have the highest refractive indices. In a series of crystals such as the alkali halides, in which all the ions have completed shells of electrons, the refractive index varies in a regular manner with the nature of the ions present. These relationships are best expressed in terms of the molecular refractivity, R, according to the formula

$$R = \frac{n^2 - 1}{n^2 + 2} \frac{M}{d},$$

where n is the refractive index, M the molecular weight, and d the density. This formula is known by the name of Lorenz-Lorentz, and the interest in expressing the refractive index, molecular weight and density in this way lies in the fact that for nearly every compound the value of R is simply the sum of the R's for each of its constituents. The expression was not originally applied to crystals but to gases, and provided the electronic configurations of the ions are not greatly altered by the juxtaposition of their neighbours in the crystal, the additive relation still holds. The molecular refractions for the alkali halides for the Na-D line are given in the following table.

TABLE XXVIII. *Molecular refractions for the alkali halides*[†]

	F	Δ	Cl	Δ	Br	Δ	I
Li	2·34	5·25	7·59	2·97	10·56	5·42	15·98
Δ	0·68		0·93		1·00		1·09
Na	3·02	5·50	8·52	3·04	11·56	5·51	17·07
Δ	2·15		2·33		2·42		2·68
K	5·17	5·68	10·85	3·13	13·98	5·77	19·75
Δ	1·58		1·70		1·80		1·96
Rb	6·74	5·81	12·55	3·23	15·78	5·93	21·71
Δ	2·77		2·70		2·68		2·56
Cs	9·51	5·74	15·25	3·21	18·46	5·81	24·27

Were the Lorenz-Lorentz relation strictly applicable the differences, Δ, between neighbouring rows and columns would be equal throughout any one row or column. It is clear that there are deviations, though generally they are less than half a unit. However, we may say that the law is valid to a first approximation and from this proceed to discuss the molecular refractions of individual ions.

If the value of the molecular refraction of one ion in the table be fixed, then the values of all the others may be found by subtraction. Values for the refractivities of ions of the same electronic (rare gas) configuration were worked out theoretically by Wasastjerna,[‡] so that the molecular refraction of any one ion in the above table was known. Further, theoretical

[†] R. Schoppe, Z. *phys. Chem.* B, **24**, 259 (1934).
[‡] J. A. Wasastjerna, Z. *phys. Chem.* **101**, 193 (1922).

studies indicate that the values for the ions of the same electronic configuration, e.g. F^-, Ne, Na^+, Mg^{++}, Al^{+++}, must lie on a smooth curve. The molecular refractivities of the rare gases can be measured directly and hence, by trial and error, it is possible to arrive at the best set of values of the ionic molecular refractions. The deviations from strict additivity are important, for it is found that the refractivity of anions is lowered by neighbouring cations, the smaller and more highly charged the cation, the greater being the effect. The refractivity of the cations is increased by neighbouring anions, and the effect increases with the field strength of the anion and with the refractivity of the cation.

An example of the importance of these considerations is the criterion which they afford for the state of packing of certain ions in some crystals. Thus in the structures in which the oxygen ions are close-packed the refractive indices (Na-D) are about 1·7,†

	BeO	1·73
Corundum	Al_2O_3	1·77
Spinel	$MgAl_2O_4$	1·72
Chrysoberyl	$BeAl_2O_4$	1·75
Cyanite	Al_2SiO_5	1·72
Topaz	$Al(FOH)_2SiO_4$	1·61

whereas in the three-dimensional framework structures of the felspars and zeolites the refractive indices are much lower, being little different from quartz, which is typical of this kind of structure.

Orthoclase	$KAlSi_3O_8$	1·52
Albite	$NaAlSi_3O_8$	1·53
Anorthite	$CaAl_2Si_2O_8$	1·58
Analcite	$NaAlSi_2O_6.H_2O$	1·49
Natrolite		1·49
Stilbite		1·50
Chabazite		1·48
Quartz	SiO_2	1·55

6.2. Ionic refractivity and ionic radius.

A relation exists between the refractivities of ions and their radii.‡ Historically this has been of great importance in the development of the study of crystal structures. It was early

† W. L. Bragg, Z. Krist. **74**, 237 (1930).

‡ J. A. Wasastjerna, Soc. Sci. Fenn. Comm. Phys. Math. **6**, 21 (1932); ibid. **1**, 38 (1923).

recognised that many crystal structures could probably be represented as assemblages of spheres in contact and it was assumed that an atom or ion had a fixed radius. By considering the cell sizes of all the alkali halides for which the side of the cell is equal to the sum of the diameters of anion and cation it is possible to draw up a table similar to that on p. 173 for the molecular refractivities. If the radius of one ion is known all the others may be determined simply by subtraction, because the radius of any one ion is substantially the same in whatever compound it occurs, provided the state of ionisation is the same. The influence of the state of ionisation was overlooked at first, and the erroneous assumption made that the radius of an atom was the same in a metal and in a salt. Wasastjerna† showed that a simple relation should exist between the refractivity of an ion having a completed electron shell and its radius, which may be expressed approximately as follows:

$$r^4 = kI,$$

where r is the radius of the ion, k is a constant not far from unity but depending on the column of the periodic table to which the ion belongs, and I is the ionic refractivity. On the basis of this relation V. M. Goldschmidt‡ founded his systematic work on crystal chemistry and in particular determined a correct set of atomic and ionic radii. The fundamental fact that the radius of an atom depends on its state of ionisation was thereby clearly established and the earlier confusion resolved.

7. The relations between the optical properties of crystals and their crystal structures.

7.1. Introduction.

The optical properties of crystals depend both on the nature of the constituent ions and on their mutual arrangement. The

† J. A. Wasastjerna, *Soc. Sci. Fenn. Comm. Phys. Math.* 6, 18, 19, 21 (1932).
‡ V. M. Goldschmidt, *Skr. norske Vidensk.-Akad.* I. Matem-naturvidensk. Klasse, 1926, Nos. 1 and 8.

influence of the refractivities of different types of ion on the refractive index of the crystal has been considered in the previous section, and in this section effects due to the mutual arrangement will be studied. According to the electromagnetic theory of light there are in an illuminated refracting medium electric and magnetic fields travelling with the velocity of light. The electric fields induce dipoles in the atoms and ions over which they pass, and the optical properties of crystals are largely determined by the mutual influence of these dipoles. This mutual influence is always considerable and, for example, in a crystal for which $\mu = 3$, the induced dipoles have a strength such that their combined reaction on the light wave reduces its velocity to one-third of that in vacuum. In contrast with this the magnetic dipoles induced by magnetic fields in diamagnetic crystals are so small that the permeability does not usually differ from unity by more than 0·0001. Thus the induced magnetic dipoles do not influence one another appreciably whereas the optically induced dipoles are much stronger and exert a considerable influence on one another. It is to be expected, therefore, and is actually the case, that much more quantitative information about the structure of crystals can be obtained from a knowledge of the magnetic properties than from a study of the optical. Optical observations can be more easily made on crystals, especially when small or difficult to handle, and there already exists much more information on the optics than on the magnetic properties of crystals. Qualitative relations between crystal structures and the corresponding optical properties have proved very useful in checking and suggesting structure types. In the next paragraph these qualitative relations are given first for non-molecular and then for molecular crystals. Molecular crystals are treated separately because the molecules behave as optical entities, and the optical properties of the crystal depend largely on the shape and mutual orientation of the molecules and but little on their constituent atoms.

7.2. Empirical relations.

A. Non-molecular crystals.

The main features of an empirical classification of the optical properties of non-molecular substances are shown in Table XXIX.† Substances are grouped according to the general nature of their lattice, or the type of ions they contain.

The double refraction of isosthenic lattices is low because ionic groups such as SO_4, ClO_4, SiO_4 behave as though almost isotropic.

The double refraction of crystals having lattices in which atoms or ions are chiefly concentrated into parallel layers is large. The sign is predominantly negative, because such layers of atoms tend to be arranged perpendicular to the principal axis of symmetry of the crystal and light travels more slowly when the vibration direction lies in these planes. The reason for this is discussed in paragraph 7.3.

Strong double refractions may also be due to ions such as CO_3, NO_3, ClO_3 which are planar or almost planar in character. If such groups are arranged parallel to one another the refractive index for vibrations parallel to the plane of the group is greater than that for a vibration perpendicular to it. In calcite and sodium nitrate this gives rise to a negative sign because the plane of the asymmetric group is perpendicular to the principal axis, but in bastnaesite the CO_3 groups are arranged parallel to the principal axis and the optic sign is therefore positive.

There are very few crystal structures which are of the chain lattice type, but those that do occur have a large double refraction. The sign is positive because such lines of atoms are generally parallel to the principal axis of symmetry and therefore the extraordinary ray travels more slowly than the ordinary ray. The linear groups N—N—N and F—H—F also give rise to strong double refraction. In NaN_3 they are arranged parallel and in KHF_2 and KN_3 perpendicular to the principal

† W. A. Wooster, Z. Krist. **80**, 495 (1931).

TABLE XXIX. *Relations between structure types and optical properties*

Isosthenic lattices†

Structure type	Substance	Refractive indices				
		α or ω	β	γ or ϵ	$\dfrac{\gamma - \alpha}{\text{or } \epsilon - \omega}$	Sign
H1, Anhydrite	$CaSO_4$	1·569	1·575	1·613	0·044	+
H0$_5$	NH_4ClO_4	1·481	1·483	1·488	0·007	+
H2, Barytes	$BaSO_4$	1·637	1·638	1·649	0·012	+
H13, Phenacite	Be_2SiO_4	1·654	—	1·670	0·016	+

Layer lattices

Structure type	Substance	Refractive indices			
		ω	ϵ	$\epsilon - \omega$	Sign
B10, PbO	PbO (red, tetragonal)	2·665 (Li)	2·535 (Li)	0·130	−
C6, CdI$_2$	CdI$_2$	—	—	Large	−
	PbI$_2$				−
D0$_4$, CrCl$_3$		—	—	>0·5	−
D0$_5$, AsI$_3$		2·59	2·23	0·36	−
G1, Calcite	$CaCO_3$	1·658	1·486	0·172	−
G0$_1$	$NaNO_3$	1·585	1·336	0·249	−
G7$_1$, Bastnaesite	(Ce, La, Di)$_2$(CO$_3$)$_3$ (Ce, La, Di)F$_3$	1·717	1·818	0·101	+
D31, Calomel	Hg_2Cl_2	1·973	2·656	0·683	+

Chain lattices

Structure type	Substance	Refractive indices			
		ω	ϵ	$\epsilon - \omega$	Sign
B9, HgS	HgS (Cinnabar)	2·854	3·201	0·347	+
A8, Se	Se	3·0	4·04	1·04	+
F51, NaN$_3$	NaN_3	—	—	Very large	+
F52, KHF$_2$	KN_3	—	—	Large	−

Lattices with three-dimensional frameworks

Structure type	Substance	Refractive indices				
		α or ω	β	γ or ϵ	$\dfrac{\gamma - \alpha}{\text{or } \epsilon - \omega}$	Sign
C8, Quartz	SiO_2	1·544	—	1·553	0·009	+
S, Felspar	$KAlSi_3O_8$	1·519	1·524	1·526	0·007	−
S, Natrolite	$Na_2Al_2Si_3O_{10} . 2H_2O$	1·475	1·479	1·489	0·014	+

† For a definition of the terms used in describing these lattice types see Chapter II, p. 34.

axis; as in the previous example the optic sign corresponds to the orientation of the asymmetric group.

In the last portion of Table XXIX there are examples of crystals having a three-dimensional framework of SiO_4 tetrahedra. The double refraction is low, because of the absence of any pronounced structural anisotropy.

Certain exceptions to these generalisations arise when ions of iron or titanium are present, but there are few of these crystals and they need not concern us here.

B. Molecular crystals.

The types of molecules which are considered below are (1) almost spherical, (2) rod-shaped, (3) planar. The double refraction depends not only on the shape of the molecules but also on their mutual arrangement. The last column of Table XXX illustrates the general rules governing the relation between the crystal structure and the corresponding double refraction.

7.3. Calculation of the double refraction of certain ionic crystals from their crystal structures.

The study of the optical properties of regular arrangements of atoms was begun before there was any exact knowledge of crystal structures. In fact, the discovery of the diffraction of X-rays by crystals was in part due to the study of the optical properties of lattices. For some time before 1912 Ewald had been studying at the institute of Sommerfeld the double refraction to be expected from arrangements of atoms on crystal lattices. The nature of X-rays was then a problem of the moment, and it is not surprising that the first experiment on the diffraction of X-rays by crystals should have been performed in the laboratory where people were studying the optics of crystal lattices as well as the nature of X-rays. In the following period, the study of crystal structures developed rapidly and there soon appeared the work of Born on the physical properties of crystals calculated on the basis of their

TABLE XXX

(1) Almost spherical molecules

Structure type	Substance	Refractive indices					Conclusion
		α or ω	β	γ or ϵ	$\gamma - \alpha$ or $\epsilon - \omega$	Sign	
O 4	Pentaerythritol, $C(CH_2OH)_4$	1·559	—	1·548	0·011	–	None

(2) Rod shaped molecules

Structure type	Substance	Refractive indices					Conclusion
		α or ω	β	γ or ϵ	$\gamma - \alpha$ or $\epsilon - \omega$	Sign	
Paraffin	$C_{29}H_{60}$† $C_{16}H_{33}OH$†	$\alpha \parallel a$ $\alpha \perp c$	$\beta \parallel b$ $\beta \parallel b$	$\gamma \parallel c$ $\gamma \parallel c$		+ +	The positive sign of the double refraction and its large magnitude are associated with structures in which the rod-like molecules are all parallel to γ
	$C_{18}H_{37}NH_3Cl$†	$\alpha \parallel b$	$\beta \parallel a$	$\gamma \parallel c$		–	The negative sign of the double refraction indicates that the lengths of the molecules must be arranged in at least two directions lying in a plane so that $\beta \approx \gamma$. α is then parallel to the normal to the plane containing the molecules
C 2	CO_2	—	—	—	0 (isotropic)	–	The isotropic or nearly isotropic character of crystals in this group shows that the lengths of the molecules must lie in at least four directions

(3) *Planar molecules with two large and one small refractive index*

Structure type	Substance	Refractive indices (NaD line)					Conclusion
		α or ω	β	γ or ϵ	$\gamma - \alpha$ or $\epsilon - \omega$	Sign	
	Naphthalene, $C_{10}H_8$‡ ($\lambda = 5461$)	1·525	1·722	1·945	0·420	+	The great value of the double refraction shows that the molecules must be at least approximately parallel
	Anthracene, $C_{14}H_{10}$‡ ($\lambda = 5461$)	1·556	1·786	1·959	0·403	−	
	Chrysene, $C_{18}H_{12}$‡ ($\lambda = 5461$)	1·585	1·787	2·068	0·483	+	
O21	Urea, $CO(NH_2)_2$	—	—	1·720	Large	+	The large positive value of the double refraction indicates that the molecules, though not parallel to one another, are all parallel to a given line
	Spiro 5:5-dihydantoin,§ $C_5H_4O_4N_4$	1·555	—	—	0·165	+	
	Resorcinol,‖ $C_6H_4(OH)_2$	1·578	1·620	1·627	0·049	−	The smallness of the double refraction indicates that the normals to the planes of the molecules must be arranged in several directions. The sign is of no special significance

† J. D. Bernal, *Z. Krist.* **83**, 153 (1932).
‡ K. S. Sundararajan, *Z. Krist.* **93**, 238 (1936).
§ Sir William Jackson Pope and J. B. Whitworth, *Proc. Roy. Soc.* A, **134**, 357 (1931).
‖ J. M. Robertson, *Proc. Roy. Soc.* A, **157**, 79 (1936).

known structures. The fundamental theory was published in 1915,[†] and in the succeeding years a number of papers appeared in which these methods were applied[‡] to calomel, quartz, rutile and anatase.

These calculations do not permit the values of the refractive indices to be determined simply from a knowledge of the constituent ions and the crystal structure—it is necessary to know the dispersion curve for at least one ray and preferably both. Although it may seem that the answer must be known before it can be calculated the comparison between theory and experiment is important, because it established the correctness of Born's fundamental assumptions and provided a valuable check on the crystal structures derived by X-rays. The structure and the dispersion curve for the ordinary ray were taken as the basis for the calculations on quartz. The dispersion curve for the extraordinary ray and also the optical rotatory power were calculated with high accuracy from this data. In the following table the observed refractive indices both for sodium and thallium light have been used in calculating the formula giving the refractive indices over a range of wave-lengths. The double refraction was not assumed but came out of the

TABLE XXXI

	ω		ϵ	
	Calculated	Observed	Calculated	Observed
Hg_2Cl_2 Calomel				
Li light	1·9555	1·9556	2·6007	2·6006
Na ,,	1·9732	1·9733	2·6568	2·6559
Tl ,,	1·9904	1·9908	2·7130	2·7129
TiO_2 Rutile				
Na light·	2·6187	2·6152	2·9258	2·9029
TiO_2 Anatase				
Na light	2·5492	2·5618	2·4933	2·4886

[†] Max Born, *Dynamik der Kristallgitter* (Teubner, Leipzig, 1915). *Atomtheorie des Festen Zustandes* (B. G. Teubner, Leipzig, 1923).

[‡] E. Hylleraas, *Phys. Z.* **26**, 811 (1925); *Z. Phys.* **36**, 859 (1926), *ibid.* **44**, 871 (1927); *Z. Krist.* **65**, 469 (1927).

application of the theory. It will be seen that the agreement between the calculated and observed refractive indices is reasonably good.

The calculations are lengthy and have not been applied to many substances, but enough has been done to show that this theory combined with a knowledge of the crystal structure and the dispersion curves for the refractive indices can predict with fair accuracy the double refraction of crystals with simple structures.

Another theory, due to W. L. Bragg,[†] was developed about 1924, and it has the advantage of simpler mathematics than that developed by Born and his co-workers. In this theory each atom is assumed to become an electric dipole under the influence of an electric field, and the double refraction is explained by the mutual reaction of neighbouring dipoles. The calculations take into account the fact that the electric field is due to light waves by using the ionic refractivities (see paragraph 6) as a measure of the polarisability of the ions. The calculations are straightforward and not very long, and interested students are recommended to consult the original papers. The method was first applied to the alkaline earth carbonates, especially calcite and aragonite, for which it was found that to a first approximation the double refraction could be explained in terms of the mutual reactions of the three oxygen ions belonging to the same CO_3 group. This may be illustrated by figs. 84 and 85, in which the long arrows passing through each oxygen of the CO_3 ion represent the electric moment induced by the electric field E. In fig. 84 the field is in the plane of the CO_3 group, and it is clear that the combined effect of ions A and C on B will be to induce an additional moment in the same direction as E. The effect of B on A will be to induce a certain moment in the direction of E and a component perpendicular to E, which will be balanced by an equal and opposite component induced in C by B. The effect

† W. L. Bragg, Proc. Roy. Soc. A, 105, 370 (1924); 106, 346 (1924). Atomic Structure of Minerals, W. L. Bragg (Oxford Univ. Press, 1937).

of C on A will be to induce a component in the opposite direction to E of magnitude less than that induced by B. From the symmetry of the figure the resultant moment of C must equal that of A. Thus the polarisation of each ion is increased by that of its neighbours. In fig. 85 the electric field is perpendicular to the plane of the CO_3 group (which is therefore drawn perpendicular to the plane of the paper). It is clear that in this case the effect of each ionic dipole is to decrease the polarisation of

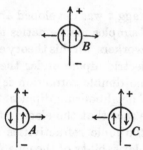

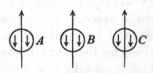

Fig. 85. The long arrows represent the dipoles induced by the field when applied perpendicular to the plane of the group. The short arrows represent the dipoles induced in the ions by one another.

Fig. 84. The long arrows represent the dipoles induced in the oxygen ions of a CO_3 group by an electric field applied in the plane of the ion. The short arrows represent the dipoles induced in the ions by one another.

each of the others, so that the resultant polarisation is less than when the electric field is parallel to the plane of the CO_3 group. The refractive index increases with the polarisation induced by the electric field and the difference between the polarisability in directions parallel and perpendicular to the principal axis corresponds to the large difference in the principal refractive indices. The influence of the calcium ions is independent of the direction of the electric field and their effect is allowed for in deriving the results given below. The influence of the carbon ions can be neglected.

The figures in the column headed '1st approximation' are derived on the assumption that only the mutual influence of oxygen ions belonging to the same CO_3 group need be considered. If the influence on a given oxygen ion of the neighbouring CO_3 ions be taken into account a second and better approximation is obtained. The same method has been applied

TABLE XXXII. *Comparison of calculated and observed refractive indices in calcite and aragonite*

Refractive indices	Calculated		Observed
	1st approximation	2nd approximation	
Calcite ϵ	1·468	1·488	1·486
ω	1·676	1·631	1·658
Aragonite α	1·503	1·538	1·530
β	1·730	1·694	1·681
γ	1·730	1·680	1·686

to sodium bicarbonate, $NaHCO_3$,[†] and the calculated refractive indices do not differ by more than 0·02 from the observed values and the orientation of the indicatrix which is predicted by the theory differs only by a few degrees from that observed. Similar results have been obtained from the four polymorphic modifications of ammonium nitrate.[‡]

7.4. Calculation of the double refractions of certain molecular crystals.

The work so far described relates only to crystals in which the polarisable ions may be supposed to have approximately spherical symmetry. In molecular crystals this is no longer the case and the different polarisability of the molecule in different directions must be taken into account. The calculation of the principal refractive indices of the crystal follows the same lines as the corresponding calculation for magnetic susceptibilities and will not be given in detail here.[§] The principal polarisabilities of the molecules and their orientation

---

† W. H. Zachariasen, *J. Chem. Phys.* **1**, 640 (1933).
‡ S. B. Hendricks, W. E. Deming and M. E. Jefferson, *Z. Krist.* **85**, 143 (1933).
§ S. B. Hendricks and W. E. Deming, *Z. Krist.* **91**, 290 (1935).

in the lattice are used, but as the influence of neighbouring molecules becomes greater, the calculated refractive indices correspond less closely with the observed values. It is assumed that the variation of the polarisability of the molecule with the direction of applied electric field can be represented by a triaxial ellipsoid. In a paper on the optical properties of a number of oxalates this method has been applied both in calculating refractive indices from the structure and in checking the X-ray determination of the molecular orientation. The difference between calculated and measured refractive indices is never more than 0·03, and the calculated and observed molecular orientations agree closely. It is therefore clear that in some well-defined molecular structures it is justifiable to attach principal molecular polarisabilities to molecules and to use these values to check the X-ray determinations of crystal structures.

There does not appear to be agreement between the results of the optical studies on the oxalates and the aromatic hydrocarbons.† For the latter the influence of the neighbouring

TABLE XXXIII

Crystal	Inclination to the c axis of the		
	length of the molecule (X-ray analysis)	γ-axis of the indicatrix	axis of maximum magnetic susceptibility
Naphthalene	12°·9	9°·6	12°·0
Anthracene	8·5	7·5	8·0
Phenanthrene	—	19·2	9·7
Fluorene	—	10·5	10·9
Chrysene	13·8	9·2	12·7
Diphenyl	20·0	30·6	20·0
Terphenyl	−15·3	−14·1	−15·3

† K. S. Sundararajan, Z. Krist. 93, 238 (1936).

molecules on the polarisability of each molecule is greater than would be expected from the work on the oxalates. Although it may not be possible to calculate the double refractions of the aromatic hydrocarbons, some information can be obtained on the molecular orientation. The vibration direction corresponding to the maximum refractive index often nearly coincides with the length of the molecule or the projection of it on a plane of symmetry. Table XXXIII compares X-ray, optical and magnetic data.

CHAPTER VI

PIEZO-ELECTRICITY

1. Introduction.

In 1880 J. and P. Curie discovered that certain hemihedral crystals developed electrical charges on their surfaces when subjected to mechanical stresses. Positive and negative charges were developed at the ends of the polar axes. The effects were first studied in the following crystals, tourmaline, calamine, cane sugar, tartaric acid, quartz, zincblende, helvine and sodium chlorate, all of which possess polar axes, but it will be shown later that piezo-electricity can occur in any crystal lacking a centre of symmetry, except those belonging to the class 432–O. The phenomena of piezo-electricity are grouped into direct and indirect effects; the direct effects include all developments of electric charge on the application of mechanical stress, and the indirect include all changes in shape due to the application of electric fields in certain directions in the crystal. The two effects are connected by the laws of thermodynamics. Piezo-electricity was used, soon after its discovery, for the production of controllable amounts of electric charge. Simply by compressing a suitably cut quartz plate between two metal plates, equal and opposite electric charges were generated on them. The piezo-electric effect was also used to measure changes of pressure when these were transient, or occurred under circumstances which made it difficult to employ other methods for their measurement. During recent years the applications of the indirect effect have been far-reaching. Alternating electric fields generated by valves used in radio technique are applied to plates of piezo-electric crystals, and the expansions and contractions of the plates which are thereby caused react on the electrical circuits. When the natural frequency of the mechanical vibration of the plate coincides with the frequency of oscillation of the electrical circuit, resonance occurs between the two. Under these circumstances much energy is imparted to the mechanical oscillations, and powerful wave trains are established in liquids surrounding the vibrating plates. The applications of these effects are numerous and will be dealt with in greater detail later; some of these important applications are the control of the frequency of radio transmitters, the detection of underwater obstructions —wrecks or submarines—the piezo-electric telephone, and the gramophone pick-up.

2.* General theory.

The general theory of components of stress and strain is given in Chapters II and VIII and the paragraph entitled 'components of stress' (p. 231) should be read before proceeding further with this chapter. The electric moment developed in the crystal, i.e. the charge multiplied by the separation of the charges, may be represented by a vector p having components p_i, and as shown in the general theory of elasticity the components of stress may be represented by a second order tensor t_{kl}. The most general linear relation that can exist between these quantities is expressed by the following equation:

$$p_i = q_{kli}t_{kl}, \qquad \ldots\ldots(1)$$

in which q_{kli} is a third order tensor transforming according to the usual rules. The predictions which the equation enables us to make about the variation of the piezo-electric effect with the direction of application of the pressure and all other similar relations are amply fulfilled by experimental measurement. The quantities q_{kli} are known as piezo-electric moduli. If instead of defining p_i in relation to the components of stress we used the corresponding components of strain we should obtain a similar relation involving the piezo-electric constants e_{kli}:

$$p_i = e_{kli}r_{kl}. \qquad \ldots\ldots(2)$$

3.* The limitations imposed on the piezo-electric moduli by the crystal symmetry.

In the absence of any symmetry the only fact reducing the number of piezo-electric moduli is that

$$t_{kl} = t_{lk},$$

which reduces the number of q's from 27 to 18. In a class of symmetry possessing a centre of symmetry all the q's vanish, because on interchanging $-X_1$, $-X_2$ and $-X_3$ for $+X_1$, $+X_2$ and $+X_3$, respectively, t_{kl} remains unchanged, while p_i changes its sign. Equation (1) can only be satisfied under these conditions if the q's are all zero, which means that piezo-electric phenomena cannot exist in crystals with a centre of symmetry.

Class 2—C_2.

The direction-cosines of the new axes X_1', X_2', X_3' relative to the old ones are given by the following scheme, the diad axis being taken parallel to X_3:

	X_1'	X_2'	X_3'
X_1	-1	0	0
X_2	0	-1	0
X_3	0	0	1

The general formula for the transformation of the q's from the old axes to the new is

$$q_{klt}' = c_{sk} c_{tl} c_{ri} q_{str}, \qquad \ldots\ldots(3)$$

as shown in Chapter I, p. 4.

The following example shows how the non-vanishing q's may be found. Inserting appropriate numbers in equation (3) we obtain:

$$
\begin{aligned}
q_{111}' = {} & c_{11} c_{11} c_{11} q_{111} + c_{11} c_{21} c_{11} q_{121} + c_{11} c_{31} c_{11} q_{131} \\
& + c_{21} c_{11} c_{11} q_{211} + c_{21} c_{21} c_{11} q_{221} + c_{21} c_{31} c_{11} q_{231} \\
& + c_{31} c_{11} c_{11} q_{311} + c_{31} c_{21} c_{11} q_{321} + c_{31} c_{31} c_{11} q_{331} \\
& + \text{all the terms in } q_{rs2} \text{ and in } q_{rs3} \\
= {} & -q_{111}, \qquad\qquad\qquad\qquad \ldots\ldots(4)
\end{aligned}
$$

since all c_{rk}'s for which $r \neq k$ vanish. If a diad axis parallel to X_3 is present the transformation we have carried out must leave the piezo-electric modulus unchanged. Thus

$$q'_{111} = q_{111}. \qquad \ldots\ldots(5)$$

Equations (4) and (5) can only be simultaneously satisfied if

$$q'_{111} = q_{111} = 0.$$

From a consideration of equation (3) it will be seen that only permutations of $c_{11}c_{22}c_{33}$, $c_{11}c_{11}c_{33}$, $c_{22}c_{22}c_{33}$, $c_{33}c_{33}c_{33}$ will be equal to unity and thereby give moduli which do not vanish on operating with the diad axis. Thus of the moduli set out in the following scheme only those marked with an asterisk are left for determination.

$$q_{111} \quad q_{121} \quad q_{131}{}^* \quad q_{112} \quad q_{122} \quad q_{132}{}^* \quad q_{113}{}^* \quad q_{123}{}^* \quad q_{133}$$
$$q_{221} \quad q_{231}{}^* \qquad q_{222} \quad q_{232}{}^* \qquad q_{223}{}^* \quad q_{233}$$
$$q_{331} \qquad\qquad q_{332}{}^* \qquad\qquad q_{333}{}^*$$

A convention is usually adopted for reducing the number of suffixes of tensors such as t_{ik} for which $t_{ik} = t_{ki}$, as shown in the following table:

Full symbols	11	22	33	23, 32	31, 13	12, 21
Reduced	1	2	3	4	5	6

Thus the scheme of piezo-electric moduli for a diad axis parallel to X_3 may be written:†

$$\begin{matrix} 0 & 0 & 0 & q_{41} & q_{51} & 0 \\ 0 & 0 & 0 & q_{42} & q_{52} & 0 \\ q_{13} & q_{23} & q_{33} & 0 & 0 & q_{63} \end{matrix}$$

† It is important to notice the difference between the abbreviated symbols used here and those of Voigt (*Lehrbuch der Kristallphysik*, p. 829). From the definition which he uses (p. 818) his constants e_{ik} are equal to e_{ki} as defined here. The relations of Voigt's piezo-electric moduli, d_{ik}, to the q_{ki} are $d_{ik} = q_{ki}$, when $i = k$, but if $i \neq k$, then $d_{ik} = 2q_{ki}$ (see also Chapter VIII).

Class m—C_s.

The plane of symmetry is supposed perpendicular to the axis X_3, so that the scheme of direction-cosines for the new axes becomes

	X_1'	X_2'	X_3'
X_1	1	0	0
X_2	0	1	0
X_3	0	0	-1

It will be noticed that in this scheme corresponding direction-cosines have just the opposite signs to those in the scheme for a diad axis. With this transformation any modulus possessing a single 3 or three 3's must vanish (in contrast with the previous case where only these moduli remained). Thus the scheme of q_{kli}'s is

$$
\begin{array}{ccccccccc}
q_{111} & q_{121} & 0 & & q_{112} & q_{122} & 0 & & 0 & 0 & q_{133} \\
 & q_{221} & 0 & & & q_{222} & 0 & & & 0 & q_{233} \\
 & & q_{331} & & & & q_{332} & & & & 0
\end{array}
$$

or, in the abbreviated form,

$$
\begin{array}{cccccc}
q_{11} & q_{21} & q_{31} & 0 & 0 & q_{61} \\
q_{21} & q_{22} & q_{32} & 0 & 0 & q_{62} \\
0 & 0 & 0 & q_{43} & q_{53} & 0
\end{array}
$$

Rhombic system.

Class 222—D_2, (V).

If in the monoclinic system the diad axis be taken parallel to X_3, only those q_{kli}'s possessing a single 3 or three 3's are not zero. If the diad axis were parallel to X_1, only those moduli with a single 1 or three 1's would not be zero, and, similarly, if the diad axis were parallel to X_2, only those moduli having a single 2 or three 2's would not be zero. Thus, in the class 222 for which all these conditions must be satisfied simultaneously only q's having all three suffixes different remain. The scheme is therefore

$$
\begin{array}{ccccccccc}
0 & 0 & 0 & & 0 & 0 & q_{132} & 0 & q_{123} & 0 \\
 & 0 & q_{231} & & & 0 & 0 & & 0 & 0 \\
 & & 0 & & & & 0 & & & 0
\end{array}
$$

or, in abbreviated form,

$$
\begin{array}{cccccc}
0 & 0 & 0 & q_{41} & 0 & 0 \\
0 & 0 & 0 & 0 & q_{52} & 0 \\
0 & 0 & 0 & 0 & 0 & q_{63}
\end{array}
$$

Class 2mm—C_{2v}.

In this class there are two vertical planes of symmetry combined with a vertical diad axis. The appropriate limitation on the q's will be obtained, therefore, if we combine a plane of symmetry perpendicular to X_1 with a diad axis parallel to X_3. The former limitation makes all q's possessing a single 1 or three 1's equal to zero and the latter limitation makes all q's possessing no 3's or two 3's zero. In the scheme below the q's are those of class 2 and those marked with an asterisk are eliminated by the plane of symmetry perpendicular to X_1.

$$
\begin{array}{ccccccccc}
0 & 0 & q_{131} & 0 & 0 & q_{132}{}^{*} & q_{113} & q_{123}{}^{*} & 0 \\
 & 0 & q_{231}{}^{*} & & 0 & q_{232} & & q_{223} & 0 \\
 & & 0 & & & 0 & & & q_{333}
\end{array}
$$

The following is therefore the scheme for this class in abbreviated symbols:

$$
\begin{array}{cccccc}
0 & 0 & 0 & 0 & q_{51} & 0 \\
0 & 0 & 0 & q_{42} & 0 & 0 \\
q_{13} & q_{23} & q_{33} & 0 & 0 & 0
\end{array}
$$

Tetragonal system.

Class 4—C_4.

The direction-cosine scheme for the new axes relative to the old is

	X_1'	X_2'	X_3'
X_1	0	-1	0
X_2	1	0	0
X_3	0	0	1

The only products of c's which will be non-zero will be those containing c_{12}, c_{21} and c_{33}. Since two rotations of 90° are equal to one of 180° the same restrictions must apply to the moduli in this system as apply in the monoclinic system and therefore we need only consider the eight non-vanishing moduli in that system.

$$q'_{131} = c_{21}c_{33}c_{21}q_{232} = q_{232} = q_{131},$$

$$q'_{231} = c_{12}c_{33}c_{21}q_{132} = -q_{132} = q_{231},$$

$$q'_{132} = q_{132} = -q_{231},$$

$$q'_{232} = q_{232} = q_{131},$$

$$q'_{113} = c_{21}c_{21}c_{33}q_{223} = q_{223} = q_{113},$$

$$q'_{123} = c_{21}c_{12}c_{33}q_{213} = -q_{213} = -q_{123} = q_{123} = 0,$$

$$q'_{223} = q_{223} = q_{113},$$

$$q'_{333} = c_{33}c_{33}c_{33}q_{333} = q_{333}.$$

Thus the scheme of moduli may be written

$$\begin{matrix} 0 & 0 & q_{131} & 0 & 0 & -q_{231} & 0_{113} & 0 & 0 \\ 0 & q_{231} & 0 & q_{131} & & & q_{113} & 0 & \\ 0 & & & 0 & & & & & q_{333} \end{matrix}$$

or, in the abbreviated form,

$$\begin{matrix} 0 & 0 & 0 & q_{41} & q_{51} & 0 \\ 0 & 0 & 0 & q_{51} & -q_{41} & 0 \\ q_{13} & q_{13} & q_{33} & 0 & 0 & 0 \end{matrix}$$

Class 42—D_4.

The introduction of a horizontal diad axis parallel to X_1 eliminates those q_{kli}'s possessing no 1's or two 1's. Therefore the moduli of class 4 are reduced to the following scheme:

$$\begin{matrix} 0 & 0 & 0 & 0 & 0 & -q_{231} & 0 & 0 & 0 \\ 0 & q_{231} & & 0 & 0 & & 0 & 0 & \\ 0 & & & 0 & & & & & 0 \end{matrix}$$

or, otherwise expressed,

$$\begin{matrix} 0 & 0 & 0 & q_{41} & 0 & 0 \\ 0 & 0 & 0 & 0 & -q_{41} & 0 \\ 0 & 0 & 0 & 0 & 0 & 0 \end{matrix}$$

Class 4*mm—C*$_{4v}$.

The vertical plane of symmetry is supposed to be perpendicular to X_1, when all the q_{kli}'s possessing a single 1 or three 1's become equal to zero. Thus the scheme of moduli becomes

$$
\begin{array}{ccccccccc}
0 & 0 & q_{131} & 0 & 0 & 0 & q_{113} & 0 & 0 \\
 & 0 & 0 & & 0 & q_{131} & & q_{113} & 0 \\
 & & 0 & & & 0 & & & q_{333}
\end{array}
$$

or

$$
\begin{array}{cccccc}
0 & 0 & 0 & 0 & q_{51} & 0 \\
0 & 0 & 0 & q_{51} & 0 & 0 \\
q_{13} & q_{13} & q_{33} & 0 & 0 & 0
\end{array}
$$

Class $\overline{4}$—*S*$_4$.

In addition to the limitations imposed by a diad axis there are also in this class those corresponding to a rotation of the axes through 90° and an inversion through a centre of symmetry. The direction-cosine scheme is therefore

$$
\begin{array}{cccc}
 & X_1' & X_2' & X_3' \\
X_1 & 0 & 1 & 0 \\
X_2 & -1 & 0 & 0 \\
X_3 & 0 & 0 & -1
\end{array}
$$

The eight q_{kli}'s characteristic of class 2 accordingly transform as follows:

$$q_{131}' = c_{21}c_{33}c_{21}q_{232} = -q_{232} = q_{131},$$
$$q_{231}' = c_{12}c_{33}c_{21}q_{132} = q_{132} = q_{231},$$
$$q_{132}' = q_{132} = q_{231},$$
$$q_{232}' = q_{232} = -q_{131},$$
$$q_{113}' = c_{21}c_{21}c_{33}q_{223} = -q_{223} = q_{113},$$
$$q_{123}' = c_{21}c_{12}c_{33}q_{213} = q_{213} = q_{123},$$
$$q_{223}' = q_{223} = -q_{113},$$
$$q_{333}' = c_{33}c_{33}c_{33}q_{333} = -q_{333} = q_{333} = 0.$$

Thus the scheme of moduli is

$$
\begin{array}{ccccccccc}
0 & 0 & q_{131} & 0 & 0 & q_{231} & q_{113} & q_{123} & 0 \\
 & 0 & q_{231} & 0 & -q_{131} & & -q_{113} & & 0 \\
 & 0 & & & 0 & & & & 0
\end{array}
$$

or, in abbreviated form,

$$
\begin{array}{cccccc}
0 & 0 & 0 & q_{41} & q_{51} & 0 \\
0 & 0 & 0 & -q_{51} & q_{41} & 0 \\
q_{13} & -q_{13} & 0 & 0 & 0 & q_{63}
\end{array}
$$

Class $\overline{4}2m$—$D_{2d}(V_d)$.

To the above limitations on the q_{kli}'s we must add those corresponding to a plane of symmetry perpendicular to X_1. Then all q_{kli}'s possessing a single 1 or three 1's become zero. The scheme of moduli becomes

$$
\begin{array}{ccccccccc}
0 & 0 & q_{131} & 0 & 0 & 0 & q_{113} & 0 & 0 \\
 & 0 & 0 & & 0 & -q_{131} & & -q_{113} & 0 \\
 & 0 & & & 0 & & & & 0
\end{array}
$$

or, in abbreviated form,

$$
\begin{array}{cccccc}
0 & 0 & 0 & 0 & q_{51} & 0 \\
0 & 0 & 0 & -q_{51} & 0 & 0 \\
q_{13} & -q_{13} & 0 & 0 & 0 & 0
\end{array}
$$

It is important to notice that this class can be equally well orientated with the diad axis parallel to X_1, which is equivalent to rotating the X_1, X_2 axes about X_3 through 45° from the position employed in deriving the above moduli. If we use this latter orientation the zero moduli are those possessing no 1's or two 1's in the q_{kli}'s. The moduli are therefore the same as those of class $\overline{4}$ except that the non-vanishing moduli of the class $\overline{4}2m$ with the former orientation are put equal to zero. Thus for this orientation the moduli are

$$
\begin{array}{cccccc}
0 & 0 & 0 & q_{41} & 0 & 0 \\
0 & 0 & 0 & 0 & q_{41} & 0 \\
0 & 0 & 0 & 0 & 0 & q_{63}
\end{array}
$$

Cubic system.

Class 432–O.

In the cubic system the symmetry requires that a cyclical interchange of the axes shall be possible without affecting the magnitude of the physical constants. Thus $q_{231} = q_{312} = q_{123}$, and so on for all the q_{kli}'s. Consider the non-vanishing moduli for class 4 and apply the above requirement. Then

$$q_{131} = q_{212} = q_{323},$$

but

$$q_{212} = q_{122} = 0;$$

hence

$$q_{131} = 0.$$

Also

$$q_{231} = q_{312} = q_{123} = 0,$$

$$q_{113} = q_{221} = q_{332} = 0,$$

$$q_{333} = q_{111} = q_{222} = 0.$$

It follows that all the moduli are zero in this class of symmetry even though it lacks a centre of symmetry.

Class $\bar{4}3m$—T_d.

The requirement of cyclical interchange can now be applied to class $\bar{4}$:

$$q_{131} = q_{212} = q_{323} = 0,$$

$$q_{231} = q_{312} = q_{123},$$

$$q_{113} = q_{221} = q_{332} = 0.$$

Hence the only non-vanishing modulus is q_{231} and the scheme expressed in abbreviated symbols is

$$
\begin{array}{cccccc}
0 & 0 & 0 & q_{41} & 0 & 0 \\
0 & 0 & 0 & 0 & q_{41} & 0 \\
0 & 0 & 0 & 0 & 0 & q_{41}
\end{array}
$$

Class 23—*T*.

If we apply the cyclical interchange of suffixes to the non-vanishing q_{kli}'s of class 2 we obtain the following results:

$$q_{131} = q_{212} = q_{323} = 0,$$
$$q_{231} = q_{312} = q_{123},$$
$$q_{232} = q_{313} = q_{121} = 0,$$
$$q_{113} = q_{221} = q_{332} = 0,$$
$$q_{223} = q_{331} = q_{112} = 0,$$
$$q_{333} = q_{111} = q_{222} = 0.$$

Hence only q_{231} is non-vanishing and we have the same scheme as for class $\bar{4}3m$.

Trigonal and hexagonal systems.

The study of the relations between the q_{kli}'s in these systems is more difficult than in the others, because the direction-cosines of the new axes relative to the old are not ± 1 or 0. For the sake of generality we shall not specify the angle of rotation, except as ϕ until later. Thus we shall work out the relations between the q_{kli}'s which hold for any value of ϕ other than 90° or 180°. The scheme of direction-cosines is

	X_1'	X_2'	X_3'
X_1	c	$-s$	0
X_2	s	c	0
X_3	0	0	1

where $c = \cos\phi$, and $s = \sin\phi$. On transforming q_{111}' we obtain

$$q_{111}' = c_{11}c_{11}c_{11}q_{111} + c_{11}c_{21}c_{11}q_{121} +$$
$$c_{21}c_{11}c_{11}q_{211} + c_{21}c_{21}c_{11}q_{221} +$$
$$c_{11}c_{11}c_{21}q_{112} + c_{11}c_{21}c_{21}q_{122} +$$
$$c_{21}c_{11}c_{21}q_{212} + c_{21}c_{21}c_{21}q_{222} +$$
$$= c^3 q_{111} + c^2 s q_{121} + c^2 s q_{211} + cs^2 q_{221} +$$
$$c^2 s q_{112} + cs^2 q_{122} + cs^2 q_{212} + s^3 q_{222}$$
$$= q_{111}.$$

Hence
$$(c^3-1)\,q_{111}+2c^2sq_{121}+cs^2q_{221}$$
$$+c^2sq_{112}+2cs^2q_{122}+s^3q_{222}=0. \qquad \ldots\ldots(6)$$

Operating in the same way on q'_{121}, q'_{221}, q'_{112}, q'_{122} and q'_{222} we obtain altogether six simultaneous equations as shown in the following scheme where each trigonometrical term is to be multiplied by the q_{kli} at the head of the column in which it is:

q_{111}	q_{121}	q_{221}	q_{112}	q_{122}	q_{222}		
$(c^3-1)+$	$2c^2s$	$+cs^2$	$+c^2s$	$+2cs^2$	$+s^3$	$=0$	$\ldots\ldots(7)$
$-c^2s$	$+(c^3-cs^2-1)+$	c^2s	$-cs^2$	$+(c^2s-s^3)$	$+cs^2$	$=0$	$\ldots\ldots(8)$
cs^2	$-2c^2s$	$+(c^3-1)+$	s^3	$-2cs^2$	$+c^2s$	$=0$	$\ldots\ldots(9)$
$-c^2s$	$-2cs^2$	$-s^3$	$+(c^3-1)+$	$2c^2s$	$+cs^2$	$=0$	$\ldots\ldots(10)$
cs^2	$-(c^2s-s^3)$	$-cs^2$	$-c^2s$	$+(c^3-cs^2-1)+$	c^2s	$=0$	$\ldots\ldots(11)$
$-s^3$	$+2cs^2$	$-c^2s$	$+cs^2$	$-2c^2s$	$+(c^3-1)$	$=0$	$\ldots\ldots(12)$

The values of these six q_{kli}'s can be found by any of the standard methods of solving simultaneous linear equations. A simple procedure is as follows: Add equations (7) and (9),

$$(c^3+cs^2-1)\,q_{111}+(c^3+cs^2-1)\,q_{221}+(c^2s+s^3)\,q_{112}+$$
$$(c^2s+s^3)\,q_{222}=0$$

or $\quad(q_{111}+q_{221})\,(c^3+cs^2-1)+(q_{112}+q_{222})\,(c^2s+s^3)=0.$

Add equations (10) and (12) and we obtain

$$-(q_{111}+q_{221})\,(c^2s+s^3)+(q_{112}+q_{222})\,(c^3+cs^2-1)=0.$$

These relations can only be satisfied if

$$q_{111}=-q_{221}; \quad q_{112}=-q_{222}. \qquad \ldots\ldots(13)$$

Similarly, on subtracting equations (8) and (10) and comparing the result with that obtained on subtracting equations (9) and (11), we obtain

$$q_{121}=q_{112}; \quad \text{and} \quad q_{221}=q_{122}. \qquad \ldots\vdots\ldots(14)$$

On combining this result with equation (13) we have

$$q_{111}=-q_{221}=-q_{122}; \quad q_{222}=-q_{121}=-q_{112}. \ldots(15)$$

Inserting the relations of equation (15) in equation (7) we obtain

$$q_{111}(c^3 - 3cs^2 - 1) + q_{222}(s^3 - 3c^2s) = 0. \quad \ldots\ldots(16)$$

Equation (15) applies to both the trigonal and the hexagonal systems, but equation (16) differentiates between them.

For the *trigonal* system, the angle of rotation in transforming the axes is 120°, and hence

$$c = -\tfrac{1}{2}; \quad s = \frac{\sqrt{3}}{2},$$

and equation (16) becomes

$$q_{111} \cdot 0 + q_{222} \cdot 0 = 0,$$

or in other words q_{111} and q_{222} may have any value and still satisfy the symmetry requirements.

For the *hexagonal* system, however, the corresponding angle of rotation is 60°, when

$$c = \tfrac{1}{2}; \quad s = \frac{\sqrt{3}}{2},$$

and equation (16) becomes

$$q_{111} \cdot 2 + q_{222} \cdot 0 = 0,$$

i.e. $\qquad\qquad\qquad q_{111} = 0.$

On inserting the relations of equation (15) in equation (8) we could, similarly, show that in this system

$$q_{222} = 0.$$

Hence the following relation applies to the hexagonal system but not to the trigonal:

$$q_{111} = q_{121} = q_{221} = q_{112} = q_{122} = q_{222} = 0. \quad \ldots\ldots(17)$$

Proceeding in the same way we obtain the following general results for both systems.

By transforming q'_{131} and q'_{231} we obtain

$$q_{131} = q_{232} \quad \text{and} \quad q_{231} = -q_{132}. \quad \ldots\ldots(18)$$

Further, $\qquad q'_{331} = c_{33}c_{33}c_{11}q_{331} + c_{33}c_{33}c_{21}q_{332}$

$$= cq_{331} + sq_{332} = q_{331}.$$

Also $\quad q'_{332} = c_{33}c_{33}c_{12}q_{331} + c_{33}c_{33}c_{22}q_{332}$

$$= -sq_{331} + cq_{332} = q_{332}.$$

This is impossible unless

$$q_{331} = q_{332} = 0. \qquad \ldots(19)$$

Also from the transformation of q'_{113}, q'_{123},

$$q_{123} = 0,$$
$$q_{113} = q_{223}. \qquad \ldots(20)$$

Finally, by transforming q'_{133} and q'_{233} we obtain

$$q_{133} = q_{233} = 0. \qquad \ldots(21)$$

Thus the scheme of moduli for the two systems differ only in the application of equation (17).

Trigonal system.

Class 3—C_3.

The following is the scheme of moduli obtained from equations (15), (18), (19), (20) and (21):

$$
\begin{array}{ccccccccc}
q_{111} & -q_{222} & q_{131} & -q_{222} & -q_{111} & -q_{231} & q_{113} & 0 & 0 \\
& -q_{111} & q_{231} & & q_{222} & q_{131} & & q_{113} & 0 \\
& & 0 & & & 0 & & & q_{333}
\end{array}
$$

$$
\begin{array}{cccccc}
q_{11} & -q_{11} & 0 & q_{41} & q_{51} & -q_{22} \\
-q_{22} & q_{22} & 0 & q_{51} & -q_{41} & -q_{11} \\
q_{13} & q_{13} & q_{33} & 0 & 0 & 0
\end{array}
$$

Class 32—D_3.

If the X_1 axis be made the diad axis then only the q_{kli}'s having one or three 1's remain non-zero. Hence the scheme becomes

$$
\begin{array}{ccccccccc}
q_{111} & 0 & 0 & 0 & -q_{111} & -q_{231} & 0 & 0 & 0 \\
& -q_{111} & q_{231} & & 0 & 0 & & 0 & 0 \\
& & 0 & & & 0 & & & 0
\end{array}
$$

$$
\begin{array}{cccccc}
q_{11} & -q_{11} & 0 & q_{41} & 0 & 0 \\
0 & 0 & 0 & 0 & -q_{41} & -q_{11} \\
0 & 0 & 0 & 0 & 0 & 0
\end{array}
$$

Class 3m—C_{3v}.

(a) Plane of symmetry perpendicular to X_1: all moduli which are non-vanishing must have two 1's or no 1's:

$$
\begin{array}{ccccccccc}
0 & -q_{222} & q_{131} & -q_{222} & 0 & 0 & q_{113} & 0 & 0 \\
0 & 0 & & q_{222} & q_{131} & & q_{113} & 0 \\
0 & & & 0 & & & & q_{333}
\end{array}
$$

$$
\begin{array}{cccccc}
0 & 0 & 0 & 0 & q_{51} & -q_{22} \\
-q_{22} & q_{22} & 0 & q_{51} & 0 & 0 \\
q_{13} & q_{13} & q_{33} & 0 & 0 & 0
\end{array}
$$

(b) Plane of symmetry perpendicular to X_2; all moduli which are non-vanishing must have two 2's or no 2's:

$$
\begin{array}{ccccccccc}
q_{111} & 0 & q_{131} & 0 & -q_{111} & 0 & q_{113} & 0 & 0 \\
& -q_{111} & 0 & & 0 & q_{131} & & q_{113} & 0 \\
& & 0 & & & 0 & & & q_{333}
\end{array}
$$

$$
\begin{array}{cccccc}
q_{11} & -q_{11} & 0 & 0 & q_{51} & 0 \\
0 & 0 & 0 & q_{51} & 0 & -q_{11} \\
q_{13} & q_{13} & q_{33} & 0 & 0 & 0
\end{array}
$$

Hexagonal system.

Class 6—C_6.

The scheme of moduli obtained from equations (17), (18), (19), (20) and (21) is

$$
\begin{array}{ccccccccc}
0 & 0 & q_{131} & 0 & 0 & -q_{231} & q_{113} & 0 & 0 \\
& 0 & q_{231} & 0 & q_{131} & & & q_{113} & 0 \\
& & 0 & & & 0 & & & q_{333}
\end{array}
$$

$$
\begin{array}{cccccc}
0 & 0 & 0 & q_{41} & q_{51} & 0 \\
0 & 0 & 0 & q_{51} & -q_{41} & 0 \\
q_{13} & q_{13} & q_{33} & 0 & 0 & 0
\end{array}
$$

It should be noted that this scheme is the same as that for class 4, which is an example of the general relation that for an nth order tensor the same scheme of coefficients applies to classes of symmetry having a principal axis of degree $(n+1)$ or higher.

Classes 62—D_6 and 6mm—C_{6v}.

From the study of class 6 it will be seen that the schemes for the other hexagonal classes must be the same as those for the corresponding tetragonal ones, viz. 42, and 4mm.

Class $\bar{6}$—C_{3h}.

The symmetry of this class may be regarded as that of a triad axis perpendicular to a plane of symmetry. Therefore of the moduli for class 3 only those having two 3's or no 3's will be non-zero. The scheme is

$$
\begin{array}{ccccccccc}
q_{111} & -q_{222} & 0 & -q_{222} & -q_{111} & 0 & 0 & 0 & 0 \\
 & -q_{111} & 0 & & q_{222} & 0 & & 0 & 0 \\
 & & 0 & & & 0 & & & 0 \\
\end{array}
$$

$$
\begin{array}{cccccc}
q_{11} & -q_{11} & 0 & 0 & 0 & -q_{22} \\
-q_{22} & q_{22} & 0 & 0 & 0 & -q_{11} \\
0 & 0 & 0 & 0 & 0 & 0 \\
\end{array}
$$

Class $\bar{6}2m$—D_{3h}.

The symmetry of this class is that of $\bar{6}$, to which a plane of symmetry parallel to X_3 has been added. In the scheme of moduli given below, the plane of symmetry is supposed perpendicular to X_2, so that the non-zero terms have two 2's or none:

$$
\begin{array}{ccccccccc}
q_{111} & 0 & 0 & 0 & -q_{111} & 0 & 0 & 0 & 0 \\
 & -q_{111} & 0 & & 0 & 0 & & 0 & 0 \\
 & & 0 & & & 0 & & & 0 \\
\end{array}
$$

$$
\begin{array}{cccccc}
q_{11} & -q_{11} & 0 & 0 & 0 & 0 \\
0 & 0 & 0 & 0 & 0 & -q_{11} \\
0 & 0 & 0 & 0 & 0 & 0 \\
\end{array}
$$

4. Experimental methods of measuring piezo-electric moduli.

Piezo-electric moduli may be measured in two ways. The static method was used in most of the standard measurements, but important though limited applications have been made of the dynamic method.

4.1. Static method.

A suitably cut plate P of a piezo-electric crystal is placed between two metal electrodes Q, which are connected to opposite pairs of quadrants of an electrometer E. On compression, the crystal plate develops a charge, and the electrometer registers a change in potential. This charge is measured

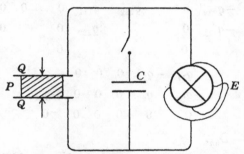

Fig. 86. Apparatus used for determining the charge liberated on the surfaces of a piezo-electric plate when under static compression.

by inserting a standard condenser, C, in parallel with the plates Q. Without C the capacity of the insulated system is equal to c_1. With C inserted, it increases to c_2. If the deflection of the electrometer for a given load on the crystal is e_1 and this changes on insertion of C to e_2, then, p being the charge developed,

$$p = c_1 e_1 = c_2 e_2.$$

If c_0 is the capacity of the standard condenser,

$$c_2 = c_0 + c_1.$$

Hence
$$c_1 e_1 = c_0 e_2 + c_1 e_2,$$

or
$$c_1 = c_0 \frac{e_2}{e_1 - e_2}$$

and
$$p = c_1 e_1 = c_0 \frac{e_1 e_2}{e_1 - e_2}.$$

If t is the total force applied, the piezo-electric modulus q is given by
$$q = \frac{p}{t} = \frac{c_0}{t} \frac{e_1 e_2}{e_1 - e_2}.$$

The moduli are usually very small so that a sensitive electrometer must be used. The greatest experimental difficulty is due to the leakage of the electric charge from the insulated electrode. This can be minimised by drying the air surrounding it, as most of the leakage usually occurs over the surface of the crystal plate.

4.2. Dynamic method.

The frequency of the natural oscillation of a quartz plate depends on the elastic modulus for extension in the direction of vibration, the density of the material, and the thickness of the plate. Thus if the elastic modulus is s, the density ρ, and the thickness t, the velocity of propagation of longitudinal vibrations, V, is given by
$$V = \sqrt{\frac{s}{\rho}},$$

and if the plate is vibrating in its fundamental mode so that its thickness is $\lambda/2$, we have
$$V = n\lambda = 2nt,$$

where n is the frequency of vibration. Thus
$$n = \frac{1}{2t} \sqrt{\frac{s}{\rho}}.$$

This then is the relation which determines the *frequency* of vibration. The *amplitude* depends upon other factors, which will be dealt with below.

It was mentioned in paragraph 1 that a quartz plate will expand or contract along the diad axis when an electric field is applied in that direction. If the electric field be alternating, the plate expands or contracts with the frequency of the field. Should this frequency be equal to the natural frequency of vibration, resonance is set up. Such frequencies are generally of the same order as those used in radio transmission; hence the apparatus used for generating these rapid alternations may conveniently be the same as that used for generating radio waves. A quartz plate between two metal electrodes reacts in a complicated manner on the current and potential flowing in the circuit in which it is inserted. At frequencies well removed from a resonance frequency the quartz behaves like a normal dielectric, and the condenser formed of it and its electrodes has a normal value. If such a condenser is

placed in parallel with the variable condenser of an oscillatory circuit containing inductance and capacity, the relevant piezo-electric modulus may be found in the following way. A current, of frequency slightly different from that corresponding to resonance, is induced† in the inductance coil and the capacity of the condenser is varied so as to make the current through the coil a maximum. This capacity is abnormally large or small, depending on whether the impressed frequency is below or above the resonance frequency (see fig. 87). The departure of the capacity of the variable condenser

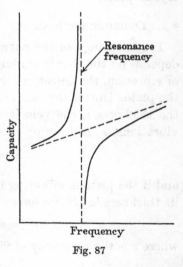

Fig. 87

† This is achieved by using a second oscillating circuit loosely coupled with the first and so designed that its frequency depends but little on the current flowing in the coil of the first circuit or on the natural frequency of oscillation of the latter.

from the normal value for a given difference between the impressed and resonance frequencies depends on the piezo-electric and elastic moduli of the quartz and its density. Since the last two quantities are known the piezo-electric modulus may be found.†‡

4.3. Qualitative methods of detecting piezo-electricity. Giebe and Scheibe method.

The reaction between an electrical circuit and a piezo-electric resonator affords a sensitive method of detecting piezo-electricity. Either a suitably cut crystal plate or a few fragments of the crystal may be used. There are several variants of the circuit which may be employed; one of the latest types is shown in fig. 88, all batteries being omitted from the diagram. P is a condenser consisting of two metal plates, with the crystal plate or fragments placed in between. Q is a variable condenser by means of which the frequency of oscillation of the circuit may be changed so as to pass through the resonant frequencies. The rest of the circuit represents an amplifier which registers on the loud speaker the changes which occur in the current through the inductance coils of the oscillator. The duplication of the circuits reduces the effects due to stray electric and magnetic disturbances. The test is carried out as follows: The plate or fragments of crystal are placed in P and

† Berta Nussbaum, *Zeit. f. Physik.* **78**, 781 (1932). The difference between the observed capacity at a frequency ν and that to be expected from the course of the capacity-frequency curve far from the resonance frequency, ν_0, is denoted by ΔC. For quartz the relation between ΔC and $(\nu_0 - \nu)$ is given by the equation

$$\Delta C.(\nu_0 - \nu) = \frac{4q^2_{11}}{s_{11}\pi\sqrt{\rho s_{11}}} \cdot \frac{o}{e},$$

where q_{11} is the piezo-electric modulus giving the relation between the charge produced on the surfaces of the crystal parallel to the condenser plates by a force normal to them, s_{11} is the elasticity modulus relating the compression normal to the plate produced by a force applied in that direction (see Ch. VIII, p. 234), ρ is the density, o and e are the semi-lengths of the block parallel respectively to the optic and electric axes.

‡ An account of the application of this method to the measurement of the change of the piezo-electric constant with temperature is given by Vigoureux in *Quartz Resonators and Oscillators*, p. 165. (H.M. Stationery Office, London, 1931.)

the vanes in Q are slowly rotated. When the substance is
piezo-electric sounds emanate from the loud speaker. These
sounds are produced by the sudden change of the current
through the condenser P whenever the frequency of the

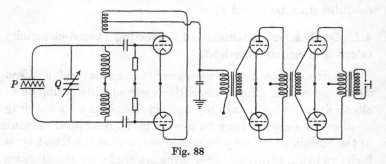

Fig. 88

oscillator corresponds to that of the crystal. With a crystal
plate a few loud cracks are heard; if there are small and
numerous grains, there is a rustling sound, which is due to a
succession of clicks. A circuit with the amplification shown in
the figure normally emits an amount of noise easy to hear, but
comfortable to listen to, but if Rochelle salt is the crystal used,
the noise is quite deafening.

4.4. Method using smoke particles.

Much of the early work on piezo-electricity was carried out
by first applying stress to certain crystals, dissipating in a
flame the charges produced and then releasing the stress so
that charges of opposite sign to those first produced were
obtained. The sign and position on the crystal of these charges
were investigated by using one or other of the smokes described
in Chapter VII, p. 224.

5.* Determination of principal piezo-electric moduli.

5.1. Calculation of charge developed on plates with an arbitrary orientation.

Rectangular or circular flat plates are usually chosen for
experiments: the use of the former will be considered here. In
general, a compression on one pair of parallel faces will produce

charges on all three pairs of faces, and we may calculate the ratio of the charge developed to the force applied in the following way. The axes are transformed so that the direction of application of the force, which is supposed normal to one pair of faces, coincides with the new X'_3 axis and the axes X'_1, X'_2 have directions parallel to the other edges of the block. The relation between the new and the old axes is specified by the c_{ik} according to the same scheme as on p. 2. The theory of this transformation is given in Appendix I. From the following relation

$$t_{11} = c_{11}c_{11}t'_{11} + c_{11}c_{12}t'_{12} + c_{11}c_{13}t'_{13}$$
$$+ c_{12}c_{11}t'_{21} + c_{12}c_{12}t'_{22} + c_{12}c_{13}t'_{23}$$
$$+ c_{13}c_{11}t'_{31} + c_{13}c_{12}t'_{32} + c_{13}c_{13}t'_{33},$$

we obtain one of the stress components referred to the original axes in terms of the stress components referred to the new axes. Since X'_3 has been chosen to coincide with the direction of application of the force

$$t'_{ik} = 0, \quad \text{except when } i = k = 3.$$

Thus $t_{11} = c_{13}c_{13}t'_{33}$.

If
$$t'_{33} = \pi,$$
then
$$t_{11} = c_{13}^2\pi.$$

Similarly, we may write down the values of all the stress components referred to the old axes:

$$t_{ik} = c_{i3}c_{k3}\pi.$$

Thus the components of the electric moment per unit volume, referred to the old axes, when due account is taken of the assumption that the applied stress is one of compression, are

$$-p_1 = q_{111}t_{11} + 2q_{121}t_{12} + 2q_{131}t_{13}$$
$$+ q_{221}t_{22} + 2q_{231}t_{23}$$
$$+ q_{331}t_{33}$$
$$= \pi(q_{111}c_{13}^2 + 2q_{121}c_{13}c_{23} + 2q_{131}c_{13}c_{33}$$
$$\left. + q_{221}c_{23}^2 + 2q_{231}c_{23}c_{33} \right\}, \quad \ldots\ldots(22)$$
$$+ q_{331}c_{33}^2)$$

and similar expressions for p_2 and p_3.

We can now find the electric moment per unit volume, parallel to the three edges of the rectangular block.

If $p_l =$ the electric moment per unit volume produced in the direction of compression,

$$p_l = p_1 c_{13} + p_2 c_{23} + p_3 c_{33},$$

where c_{i3} are the direction-cosines between the direction of compression and the axes X_1, X_2, X_3. Inserting the values of p_i from equation (22) we obtain

$$
\begin{aligned}
-\frac{p_l}{\pi} = {}& q_{111} c_{13}^3 + q_{222} c_{23}^3 + q_{333} c_{33}^3 \\
& + c_{13}^2 (2q_{131} c_{33} + 2q_{121} c_{23} + q_{112} c_{23} + q_{113} c_{33}) \\
& + c_{23}^2 ((q_{223} + 2q_{232}) c_{33} + (q_{221} + 2q_{122}) c_{13}) \\
& + c_{33}^2 ((q_{331} + 2q_{133}) c_{13} + (q_{332} + 2q_{233}) c_{23}) \\
& + 2 c_{13} c_{23} c_{33} (q_{231} + q_{132} + q_{123}) \\
= {}& q'_{333}. \qquad \qquad \text{......(23)}
\end{aligned}
$$

There are two transverse electric moments denoted p_t and p_s, the expressions for which may be derived from the equations

$$p_t = p_1 c_{11} + p_2 c_{21} + p_3 c_{31} = -\pi q'_{331}, \qquad \text{......(24)}$$

$$p_s = p_1 c_{12} + p_2 c_{22} + p_3 c_{32} = -\pi q'_{332}, \qquad \text{......(25)}$$

in the same way as the expression for p_l was derived above.

So far we have calculated the electric moments in the directions of the edges of the rectangular parallelepiped, but the quantity measured is the charge developed on these faces. If e' is the charge, and f' the area of the face,

$$e' = f' p',$$

since p' has been defined as the electric moment per unit volume.

If in the crystal block f'_1, f'_2, f'_3 are the areas of the faces perpendicular to the edges X'_1, X'_2, X'_3, then

$$e'_1 = f'_1 p_t = -f'_1 \pi q'_{331},$$
$$e'_2 = f'_2 p_s = -f'_2 \pi q'_{332},$$
$$e'_3 = f'_3 p_l = -f'_3 \pi q'_{333}.$$

If the total pressure is Γ, then

$$\pi = \Gamma/f_3',$$

or

$$e_1' = -\Gamma q_{331}' \frac{f_1'}{f_3'},$$

$$e_2' = -\Gamma q_{332}' \frac{f_2'}{f_3'},$$

$$e_3' = -\Gamma q_{333}'.$$

If the values of q_{33i}' derived from equations (23)–(25) be inserted in these equations we can calculate the charges developed on the three faces.

5.2.* Application of the theory to quartz.

The piezo-electric moduli for this crystal are given in the following scheme:

$$
\begin{array}{ccccccccc}
q_{111} & 0 & 0 & 0 & -q_{111} & -q_{231} & 0 & 0 & 0 \\
 & -q_{111} & q_{231} & 0 & 0 & & 0 & 0 \\
 & & 0 & & 0 & & & 0
\end{array}
$$

hence from equation (22)

$$
\left.\begin{aligned}
-p_1 &= \pi(q_{111}c_{13}^2 - q_{111}c_{23}^2 + 2q_{231}c_{23}c_{33}) \\
-p_2 &= \pi(-2q_{111}c_{13}c_{23} - 2q_{231}c_{13}c_{33}) \\
-p_3 &= \pi(0)
\end{aligned}\right\} \quad \ldots\ldots(26)
$$

(a) If the pressure be applied in the direction of the X_1 axis

$$c_{13} = 1, \quad c_{23} = 0, \quad c_{33} = 0$$

and $\qquad -p_1 = +\pi q_{111}, \quad -p_2 = 0, \quad -p_3 = 0.$

(b) If the pressure be applied in the direction of the X_2 axis

$$c_{23} = 1, \quad c_{13} = 0, \quad c_{33} = 0$$

and $\qquad -p_1 = -\pi q_{111}, \quad -p_2 = 0, \quad -p_3 = 0.$

(c) If the pressure be applied in the direction of the X_3 axis

$$c_{13} = 0, \quad c_{23} = 0, \quad c_{33} = 1$$

and $\qquad p_1 = p_2 = p_3 = 0.$

Further, if Γ is the total pressure and e is the charge developed,

in (a)
$$e_1 = f_1 p_1 = -f_1 \pi q_{111} = -\Gamma q_{111}, \qquad \ldots\ldots(27)$$
$$e_2 = f_2 p_2 = 0,$$
$$e_3 = f_3 p_3 = 0;$$

in (b)
$$e_1 = f_1 p_1 = f_1 \pi q_{111} = f_1 \frac{\Gamma}{f_2} q_{111}, \qquad \ldots\ldots(28)$$

or, if l = length of the block in the direction perpendicular to a diad axis and perpendicular to the optic axis, t = thickness in the direction of the same diad axis,

$$e_1 = \Gamma q_{111} \frac{l}{t},$$

$$e_2 = f_2 p_2 = 0,$$

$$e_3 = f_3 p_3 = 0;$$

in (c) no charge is produced on any face.

The results expressed in equations (27) and (28), which we have derived entirely theoretically, were originally discovered empirically by the Curies.

Details of the determination of the moduli q_{111} and q_{231}.

The complete study of the piezo-electric properties of quartz was carried out on a plate a few millimetres thick cut parallel to the optic axis and perpendicular to a diad axis. From this plate rectangular parallelepipeds were cut, the orientation of the longest sides with respect to the optic axis being known. The faces of the blocks which were perpendicular to the diad axis were covered with tinfoil. The charge which collected was measured when pressure was applied to each pair of faces in turn. From the previous paragraph we see that when the pressure is applied along the diad axis

$$-p = \pi q_{111},$$

and when pressure is applied in a direction perpendicular to the diad axis and making an angle θ with the trigonal axis

$$c_{13} = 0, \quad c_{23} = \sin\theta, \quad c_{33} = \cos\theta;$$

hence
$$-p = \pi(-q_{111}\sin^2\theta + 2q_{231}\sin\theta\cos\theta).$$

The various plates yielded a series of such equations by solution of which it was found that

$$q_{111} = -6\cdot4 \times 10^{-8}, \quad 2q_{231} = 1\cdot44 \times 10^{-8} \text{ c.g.s.e.s.u.}$$

Using these values the calculated and observed moduli for any one plate agreed within 5 per cent.

6. Experimental results for quartz, tourmaline and Rochelle salt.

Quartz.

Since the pioneer work of the Curies a number of measurements have been made of the two piezo-electric moduli q_{111}, q_{231}. The values for q_{111} are quoted as $-6\cdot32 \times 10^{-8}$, $-6\cdot45$, $-6\cdot27$, $-6\cdot94$, $-6\cdot4$ and the values for $2q_{123}$, $1\cdot45 \times 10^{-8}$, $1\cdot7$, $1\cdot925$. The range of these values is probably far outside the experimental error involved in measurements in any one crystal plate. It is difficult to obtain quartz which is quite homogeneous and twinning will generally reduce the values of the piezo-electric moduli. We must therefore attribute these variations to differences in the specimens used.

The variation of the modulus q_{111} with temperature has been studied by the dynamic method, and it should be noted that owing to the increase in conductivity of quartz the static method becomes useless at temperatures not far above room temperature. The modulus q_{111} remains practically constant until 500° C. and then drops fairly quickly so that at 567° C., i.e. 6° below the transition from α- to β-quartz, it is about one-half of its value at room temperature. At 573° C. the crystal becomes β-quartz, which has the symmetry of class 622; hence the modulus q_{111} is zero above this temperature though q_{231} is not.† Thus β-quartz is piezo-electric but not in the same way as α-quartz.

Tourmaline.

Tourmaline has been very carefully studied both on account of its piezo-electric and also its pyro-electric properties (see

† H. Osterberg and J. W. Cookson, *J. Franklin Inst.* **220**, 361 (1935).

Chapter VII). All the piezo-electric moduli were found by cutting plates having appropriate orientation to the crystallographic axes and analysing the results in a manner similar to that described in paragraph 5.2 for quartz.

Rochelle salt.

The temperature variation of the modulus q_{231} is very great. Below $-16°$ C. this modulus is relatively small, but above this temperature it rises at first gradually and then very rapidly so that at $22.5°$ C. the modulus is large but not reproducible from one specimen to another. A value as great as $26,000 \times 10^{-8}$ has been recorded. Beyond $22.5°$ C. it falls sharply within a degree or so, and beyond about $30°$ C. the conductivity rises so much that the piezo-electric modulus is difficult to measure. The other two moduli, q_{132}, q_{123}, have normal values and show no remarkable change with temperature. The very large piezo-electric modulus of Rochelle salt makes it valuable for certain practical purposes, but it is sensitive to mechanical treatment, to water vapour and any one crystal may have characters which deviate considerably from the normal. The piezo-electric modulus q_{231} is greatly reduced by the substitution of ammonium for potassium, and the relationship between the proportion of ammonium to potassium in the mixed crystal to the modulus q_{231} is complicated.

7.* The representation of piezo-electric properties by geometrical figures.

Like other physical properties piezo-electricity can be represented by geometrical figures. Each component of stress is related to three components of electric moment and therefore there are many possible surfaces to choose from. Only one or two of these are of general use, and we shall only consider the derivation of the so-called 'longitudinal piezo-electric surface'. Each radius vector of this surface has a length proportional to the charge which would be produced by unit force acting in this direction on unit area of a plate cut at right angles to it.

We may derive the form of sections of the surface for quartz by applying the results of paragraphs 5 and 6. The electric moment in a direction parallel to that in which the force is applied is given by p_l, where

$$p_l = p_1 c_{13} + p_2 c_{23} + p_3 c_{23}.$$

From equation (26)

$$p_l = -\pi(q_{111} c_{13}^2 - q_{111} c_{23}^2 + 2q_{231} c_{23} c_{33}) c_{13}$$
$$- \pi(-2q_{111} c_{13} c_{23} - 2q_{231} c_{13} c_{33}) c_{23}$$
$$= -\pi q_{111}(c_{13}^2 - 3c_{23}^2) c_{13}.$$

First we shall determine the form of the intersection of this surface with the plane containing the axes X_1 and X_2.

For all radius vectors lying in this plane

$$c_{13} = \cos\theta, \quad c_{23} = \sin\theta, \quad c_{33} = 0.$$

Now $$\cos\theta(\cos^2\theta - 3\sin^2\theta) = \cos 3\theta;$$

hence $$p_l = -\pi q_{111} \cos 3\theta.$$

The variation of p_l is therefore represented in this plane by a curve similar to a clover leaf, as shown in fig. 89. The form of the section of this surface by the $X_1 X_3$ plane may be obtained by putting $c_{23} = 0$, when

$$p_l = -\pi q_{111} c_{13}^3.$$

If θ is the angle made by the radius vector of the surface with the X_1 axis, then

$$p_l = -\pi q_{111} \cos^3\theta.$$

Thus this section of the surface has the form shown in fig. 90. The whole representation surface therefore has a shape not unlike three almonds whose pointed ends meet in the trigonal axis. This form of surface is common to all crystals belonging to the class 32.

Proceeding in the same way longitudinal piezo-electric surfaces may be drawn for tourmaline, Rochelle salt, zinc-blende, etc. The student is referred to Voigt's *Lehrbuch der Kristallphysik*, §§ 429, 431, for details of these. Briefly, the surfaces for Rochelle salt, zinc-blende and sodium chlorate are essentially alike, consisting of four bodies like zeppelins, meeting in a point, their axes coinciding with the four cubic

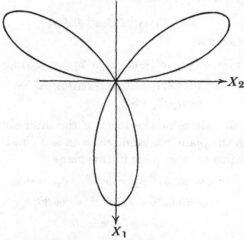

Fig. 89. A horizontal section through the origin of the longitudinal piezo-electric surface of quartz. X_1 is a diad axis. (Reproduced from *Lehrbuch der Kristallphysik*, W. Voigt (Teubner, Leipzig, 1928), by kind permission of the publishers.)

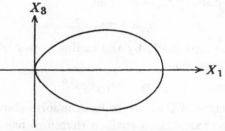

Fig. 90. A vertical section through the trigonal and diad axes of the longitudinal piezo-electric surfaces in quartz. (Reproduced from *Lehrbuch der Kristallphysik*, W. Voigt (Teubner, Leipzig, 1928), by kind permission of the publishers.)

triad axes (even though Rochelle salt is rhombic). The longitudinal piezo-electric surfaces for the classes which do not possess a polar axis, namely 1—C_1, m—C_s, 222—D_2, 422—D_4, $\bar{4}$—S_4, $\bar{4}2m$—D_{2d}, 622—D_6, $\bar{6}$—C_{3h}, $\bar{6}2m$—D_{3h}, have the following characteristics: the surfaces for 1 and m are complicated and depend on the relative values of several moduli; those for 222, $\bar{4}$, $\bar{4}2m$, though not all of identical analytical form, have four parts tetrahedrally arranged with respect to one another. The surface for $\bar{6}2m$ is the same as the quartz surface; and that for $\bar{6}$ is similar. The longitudinal piezo-electric moments for classes 422 and 622 vanish in all directions, so that there is no corresponding surface, but it is clear from the fact that these classes are piezo-electric that some of the other possible surfaces would not vanish.

8. Experimental results for other crystals.

At least four hundred substances have been examined by the Giebe and Scheibe method, described in paragraph 4.3. This is a relatively small number, but it is enough to enable some general conclusions to be drawn about the types of compounds in which piezo-electricity may be expected to occur. It should be noted that the crystals, which on morphological grounds are placed in classes of symmetry having a centre of symmetry, may be found to be piezo-electric. Conversely, crystals which have been placed in classes lacking a centre of symmetry may have no detectable piezo-electric character. In the former case it is usual to assign the crystal to a lower class than the angles between its faces require (see also Chapter VII). Where a piezo-electric effect cannot be detected, even though the external development shows a centre of symmetry to be absent, the piezo-electric observation does not affect the assignment of the crystal to its symmetry class.

The following statements give a summary of our knowledge derived from these qualitative observations:

(1) Simple ionic salts are not piezo-electric. (A simple salt is defined as one containing no radical.)

(2) Simple homopolar compounds tend to be piezo-electric, e.g. zinc-blende.

(3) Salts containing radicals of octahedral, square or tetrahedral form are not generally piezo-electric, e.g. $Fe(CN)_6$,

$PtCl_6$, SiF_6, $PtCl_4$, SO_4, ClO_4. Exceptions may occur when one of the cations is small, e.g. $LiNaSO_4$, NH_4F.

(4) Salts containing radicals which lack a centre of symmetry are often piezo-electric, e.g. ClO_3, BrO_3, IO_3, H_2PO_4, HCOO.

(5) Crystals containing symmetrical or almost symmetrical molecules are generally not piezo-electric. Thus 1-3-5-tri-nitrobenzene and ten of its substitution products are not piezo-electric.

(6) Crystals containing asymmetric molecules are often piezo-electric, e.g. $(C_6H_5)_3CH$, $CO(NH_2)_2$. Where such molecules associate in pairs the crystal is not piezo-electric.

9. Applications of piezo-electricity.

9.1. Piezo-electric voltmeter.

The contraction or expansion of a quartz plate when an electric field is applied along the diad axis may be magnified and used to measure the potential difference applied to the two sides of the plate.†

9.2. Piezo-electric pick-up.

This instrument is constructed so that the movement of the needle on the gramophone record gives rise to alternating pressures on a piezo-electric plate. The consequent alternations of potential on the electrodes attached to the plate are transmitted to the amplifier. The linear relation between pressure applied and charge produced ensures that the amplifier has an undistorted input.

9.3. Frequency standards.

The temperature coefficient of the natural frequency of vibration of a quartz plate is small. It is higher for the harmonics than for the fundamental, for which it is a decrease of five parts per million per ° C. rise in temperature. It is comparatively easy to maintain the temperature of the resonator constant to within a fraction of a degree, so that a frequency

† J. and P. Curie, *J. Physique*, 2nd series, **8**, 149 (1889).

standard of great constancy is obtainable. By suitable electric circuits piezo-electric resonators may be made to control clocks, and it is interesting to notice that a change in frequency of 1 in 10^6 corresponds to an error of 1 second in about 12 days. Probably the most important commercial application is to the maintenance of constant frequency in broadcasting transmitters. A quartz crystal may be used either to generate or to control the frequency of oscillations.

The adjustment of circuits to particular frequencies has been made particularly easy by the introduction of Giebe and Scheibe's luminous resonators. Plates or rings of quartz are mounted between suitable electrodes in an atmosphere of air or rare gas maintained at a pressure of a few millimetres of mercury. When the quartz is excited by a source of alternating potential delivering a few watts, the nodes are shown up by the gas glowing brightly in their immediate neighbourhood. The nodes are marked out in this way because they are the regions of maximum electric field where the atoms of the gas can be suitably excited. If the plate is vibrating with its fundamental frequency the bright light indicating the position of the node is at the centre of the plate; if it is vibrating in some other mode, other glowing nodes appear. It may be determined by inspection in which of its characteristic frequencies the resonator is vibrating.

The formula given on p. 205 for the natural frequency of vibration of a plate is only approximate, in that it takes no account of the dimensions and shape of the resonator. The following empirical formulae give more accurate results. If the natural frequencies of vibration along a diad axis, optic axis and an axis perpendicular to both of these are n_e, n_o, n_t, then

$n_e = 287/e$ kilocycles per sec. for discs and plates,

$n_o = 383/o$ for discs only,

$n_t = 272/t$ for discs and long thin rods cut parallel to the axis t,

where e, o, t are the thicknesses in centimetres in the corresponding directions.

9.4. Submarine signalling.

The problem of sending a directed beam of sound waves with considerable amplitude a long distance through water was solved by Langevin in 1918. The essence of Langevin's discovery was that a piezo-electric plate of large area and suitable thickness could be set into resonance and used to generate plane waves in water. The lower limit to the wave-length proved to be the heavy absorption, the upper limit the divergence of the beam; for according to the theory of diffraction most of the radiation will be contained within a cone of semi-angle l/d, where l is the wave-length and d the diameter of the generating plate, provided l is small compared with d. A wave-length of 4 cm. was found to be suitable; absorption reduced the intensity of the beam to one-third of its original value in 30 km.

The relatively enormous expense of using quartz plates 7·2 cm. thick and 20 cm. across, which were necessary to generate such radiation, was overcome by building up a plate of the required diameter from sections of quartz 2 mm. thick, cut perpendicular to a diad axis, and cementing this mosaic between two steel plates 3 cm. thick. These plates were used as the electrodes, one being in contact with the water, the other with the interior of the ship. This composite plate formed both sender of the supersonic waves and receiver of those reflected from any solid object. The distance of this object could be gauged by sending out a short train of waves at regular intervals, and registering both them and the reflected waves on the same moving paper. The interval between sending and receiving the waves was a measure of the distance they had travelled. The rate of change of this interval gave the relative velocities of the emitter and the reflector. This method has been applied to the location of all kinds of underwater obstructions, to exploration of the depth of the sea floor, and to a lesser extent to a determination of its nature.

9.5. Disintegration and emulsification produced by supersonic radiation.

A number of remarkable effects can be produced when relatively strong beams of supersonic waves are passed through liquids.† A quartz plate set into resonance at the bottom of a bath of oil produced a radiation pressure great enough to cause the surface of the liquid to form a hump in the middle of the beam and a fine spray was shot off from the hump to a considerable height. A glass rod immersed in the oil in the path of the beam transmitted the radiation along its length. A point on the glass rod set light to a small piece of wood brought into contact with it, and bored a hole through a piece of glass pressed against it. Small fish introduced into water subjected to this radiation were killed, and severe damage to human fingers was caused by taking hold of a glass rod one end of which dipped into the oil. The temperature of the glass did not rise appreciably until it was held, when it rapidly got hot where it rubbed against the fingers. When a little mercury and oil were put into a tube in the path of the beam they were rapidly emulsified. Supersonic radiation is also used to bring about a rapid 'ageing' of whiskey.

9.6. Effects consequent on change of double refraction caused by vibration.

In common with most transparent media quartz changes its double refraction when compressed or extended. One method of showing this change is as follows. A plate of quartz cut parallel to the optic axis is placed in the 45° position between crossed nicols. Light is transmitted because of the ordinary double refraction of the crystal. A similarly cut plate of quartz is arranged with its optic axis perpendicular to that of the first plate, so that the two plates compensate and no light is transmitted through the analyser. If the first plate is now set into resonance, by applying to it a suitable alternating electric potential, light is transmitted through the analyser. On examination by a rotating mirror this light is found to pulsate

† R. W. Wood and A. L. Loomis, *Phil. Mag.* 4, 417 (1927).

with twice the frequency of the mechanical vibration of the quartz. Thus it is possible to obtain a light beam pulsating with a very high frequency, and under suitable conditions this frequency is easily maintained constant to a few parts in a million. Such a beam may advantageously replace the rapidly rotating wheels or mirrors used in the classical experiments for the determination of the velocity of light. Light paths of only a few metres are required for an equally accurate determination of this velocity. This arrangement may be used to vary the intensity of a light beam in synchronism with the variation of current or potential in an electric circuit. This principle is, of course, applicable to television technique. The Herzian waves after modulation are applied to a piezo-electric quartz plate resonating with the carrier wave. The intensity of the light transmitted corresponds to the modulations of the carrier wave.

9.7. Light-scattering by oscillating crystals.

It has recently been shown† that when a crystal is set into mechanical oscillation it develops the property of diffracting a beam of light passing through it into a number of discrete beams which lie round the undiffracted beam like the spots of a Laue photograph. The patterns obtained correspond to the symmetry of the crystal in the direction in which the light travels, and from some of the patterns the elastic constants may be obtained directly. The method promises to be fruitful in obtaining the values of elastic constants of transparent crystals, and has a further advantage in that smaller crystals may be used than are required by the other methods. Similar patterns are obtained when light is reflected from crystals when they are in vibration. The detailed explanation of these patterns is complicated, but in general terms one may say that the mechanical vibration impressed on the resonating crystal sets up trains of pressure waves which, as they travel through the crystal, act as diffraction gratings which scatter the light.

† C. Schaefer, H. Bergman and L. Ludlorff, *S.B. Preuss. Akad. Wiss.* Math. Phys. Kl. p. 222 (1935).

CHAPTER VII

PYRO-ELECTRICITY

1. Introduction.

It was known two centuries ago by the traders coming from the East that heated tourmaline attracted small light objects such as pieces of paper, but its scientific study did not begin until the latter half of the last century. A tourmaline crystal has faces which give it a more pointed appearance at one end than the other. The pointed end becomes positively charged when the temperature is raised and is known as the 'analogous' pole; the other is called the 'antilogous' pole.† On cooling a crystal the surface of which has been freed from electric charges either by allowing it to stand a long time in an ordinary atmosphere, or by passing through a flame, charges are developed but in the opposite sense to that obtained on heating. The phenomena of pyro-electricity are shown by a restricted group of crystals. The feature common to them all is the absence of a centre of symmetry. Thus a study of the pyro-electric character of any crystal affords a means of determining the presence or absence of a centre of symmetry. This diagnostic test is almost the only practical application of the knowledge about pyro-electricity.

2. Methods of detecting and measuring the charge developed on pyro-electric crystals.

The qualitative methods of testing for pyro-electricity are mostly variants of a method devised by Kundt,‡ which consists in allowing charged dust particles to settle on the crystal. The most convenient way of carrying out the experiment is as follows. The crystal is heated to a suitable temperature (for

---

† One may remember that the analogous pole develops a positive charge as a result of a positive temperature change.

‡ A. Kundt, *Wied Ann.* **20**, 592 (1883).

tourmaline about 200° C.) and then passed through a flame to dissipate all the charges which have arisen on the surface. The crystal is then placed on a glass plate and allowed to cool down. An inch or so of magnesium ribbon is ignited and burnt under a bell jar, which when full of smoke is placed over the crystal. Almost at once fine filaments of magnesium oxide grow out along the lines of force and form a pattern round the crystal strongly resembling that given by iron filings around magnets.[†] The method used by Kundt consisted in shaking a mixture of powdered red lead and sulphur through a piece of muslin. During its passage through the meshes of the cloth the particles became oppositely charged, the red lead positively and the sulphur negatively. When the powder settled on the crystal there appeared red and yellow regions, indicating respectively the negatively and positively charged portions of the crystal. A third method[‡] depending on the same principle is carried out by cooling the crystal in liquid air and then suspending it freely in the room. The crystal condenses the water vapour of the surrounding air to ice particles, which form a fine powder and grow out in filaments several millimetres long along the lines of force. All these methods depend on the concentration of the electric field at the point where a dust particle has settled on the crystal. This concentration of the field causes the next particle to settle on the first rather than on some other neighbouring part of the crystal. Thus we have a growth of filaments rather than a uniform coating of powder.

Even when the pyro-electric effect of a crystal is weak it may be detected by the following method.[§] The crystal is suspended in liquid air at the end of a long glass fibre and gradually brought near to a metal plate immersed in the liquid air. If the crystal has developed charges on cooling it will be electrostatically attracted to the plate.

[†] M. E. Maurice, *Proc. Camb. Phil. Soc.* **24**, 491 (1930).
[‡] L. Bleekrode, *Ann. Phys.* **12**, 218 (1903).
[§] A. J. P. Martin, *Min. Mag.* **22**, 519 (1931).

A variant of this experiment consists in mounting a metal plate on a glass rod and holding it horizontally in liquid air. The crystals in the form of small grains are then dropped on to the plate. After the grains have cooled down, the plane of the plate is gradually brought to the vertical, when, if the grains are pyro-electric, they will cling to the plate and may even be difficult to dislodge.

3. The measurement of the pyro-electric constant.

Quantitative measurements of the charge developed on heating or cooling a crystal are difficult to make because of the readiness with which electrical charges leak away. The experiments are therefore designed so as to reduce the potentials developed and to increase the resistance to leakage. A compensation method by which the potential gradient through the crystal is always zero is due to the Curies.[†] The crystal under test and a suitably cut plate of quartz are provided with electrodes which are connected in parallel. As shown in Chapter VI, on compressing or extending the quartz charges are developed on the plates in amount proportional to the pressure applied. The pyro-electric crystal is gradually heated and the charge developed is neutralised as soon as it is liberated by applying pressure to the quartz. An electrometer is used simply as a null instrument to indicate when more charge must be neutralised. The amount of charge liberated by the quartz may be found by a subsidiary experiment as described in Chapter VI, and hence the absolute amount of charge produced by heating the pyro-electrical crystal may be found.

A method due to Gauguin[‡] employs the flicking electrometer to prevent the potential across the crystal ever rising beyond a predetermined value. A gold leaf is suspended on an insulating support so that its free end can swing and touch a metal plate. One electrode on the pyro-electric crystal is connected to the metal plate and the other to the gold leaf. As

† P. and J. Curie, *C.R. Acad. Sci., Paris*, **93**, 204 (1881).
‡ Numerous publications in *C.R. Acad. Sci., Paris* (1856–1859).

the crystal is heated equal and opposite charges flow to the gold leaf and metal plate and they are therefore attracted to one another. The gold leaf moves towards the plate and when the potential is high enough touches it. The charges are at once neutralised and the gold leaf falls back to its original position. This process is repeated so long as there is any generation of electric charge. The number of flicks of the gold leaf is a direct measure of the charge developed, and by discharging continuously the danger of loss of charge through leakage is much reduced. With this apparatus Gauguin discovered the fundamental law relating to the development of pyro-electric charge, namely, that the charge developed at either end of an electric axis depends only on the initial and final temperature of the crystal and not on the rate of heating, provided this rate be great enough to avoid loss of charge by leakage. Further, the sign of the charge, but not its magnitude, changes on interchanging the initial and final temperatures. The charge is proportional to the area of cross-section perpendicular to the electric axis but independent of the length of the crystals.

4. Electro-caloric effect.

It follows as a consequence of thermodynamics that, on applying an electric field along an electric axis of a pyro-electric crystal, there is a heating effect if the field is directed from the analogous to the antilogous pole and a cooling effect if applied in the opposite direction. The sign and magnitude of this effect have been verified† for tourmaline, though the temperature change is only $\frac{1}{500}$° C. for fields of 30,000 volts/cm.

5. True and false pyro-electricity.

The laws governing the development of pyro-electric charge in tourmaline probably apply to most but not to all substances showing pyro-electricity. For instance, these laws do not apply to quartz. It is easy to develop strong electric fields at the surface of a quartz sphere by first heating it in a spirit flame

† Fr. Lange, Diss. Jena, 1905.

and then immersing in liquid air. If the sphere is removed from the cooling bath as soon as the liquid has ceased boiling and suspended in the room six lenticular formations of ice filaments develop at the ends of the diad axes and show that they have become strongly electrified.† If the quartz be heated and then maintained at a constant temperature no such strong electrification can be observed. It is therefore necessary to establish a temperature gradient in this crystal in order to obtain a separation of the electric charges. We have seen that quartz is piezo-electric and therefore any stresses caused by temperature gradients would give rise to developments of charges at the ends of the electric axes. When the temperature of the crystal becomes uniform the stresses are much reduced (though not zero) and hence the charges also become much less. The pyro-electricity of quartz is therefore a secondary phenomenon consequent on the strains caused by rapid heating or cooling. The same would be expected to be true of all crystals which lack a centre of symmetry but do not possess a single polar axis. The pyro-electricity developed in this way is sometimes called 'false pyro-electricity' in contradistinction to 'true pyro-electricity' shown by crystals having one and only one polar axis.

The terms 'true' and 'false' are misleading, for there is no distinction between them which can be readily established. Even for tourmaline it is difficult to establish that a considerable fraction of the pyro-electric effect is due to 'true' pyro-electricity. The separation‡ of the 'true' effect from the 'false' depends on a complicated theory and a knowledge of many physical constants of the crystal.

Confusion has sometimes arisen because of the statement that 'true' pyro-electricity can only be shown by crystals having a single polar axis. An observation of pyro-electricity has been taken to prove the presence of such an axis. This may

---

† After carrying out this experiment many times the quartz sphere often breaks.

‡ W. C. Röntgen, *Ann. Phys.* **45**, 737 (1914); W. Voigt, *ibid.* **46**, 221 (1915).

be incorrect, since in practice 'true' and 'false' pyro-electricity cannot conveniently be distinguished. It is therefore safer to assume that if pyro-electricity is developed, no matter what the conditions of heating or cooling, then the crystal lacks a centre of symmetry.

6. Examples of pyro-electric crystals.

The following examples of pyro-electric crystals are chosen from several crystal systems.

Cane Sugar ($C_{12}H_{22}O_{11}$). This crystal belongs to class 2 of the monoclinic system and fig. 91 shows the difference in

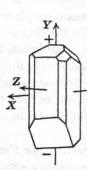
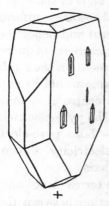

Fig. 91. Drawing of a crystal of cane sugar showing the signs and positions of the charges developed on heating. (Reproduced from *Lehrbuch der Kristallphysik*, W. Voigt (Teubner, Leipzig, 1928), by kind permission of the publishers.)

Fig. 92. Drawing of a crystal of calamine showing the signs and positions of the charges developed on heating. (Reproduced from *Crystallography and Practical Crystal Measurement*, A. E. H. Tutton (Macmillan, London, 1922), by kind permission of the publishers.)

the development of the two ends. The positive and negative signs indicate the signs of the charges which are observed on heating and the vertical arrow the direction of the diad axis.

Calamine (hydrated zinc silicate). This crystal belongs to class *mm* of the rhombic system and therefore has a single diad axis.

Pentaerythritol (C(CH₂OH)₄). This crystal belongs to class 4mm of the tetragonal system and has a unique tetrad axis.

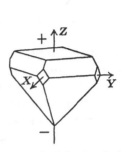

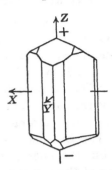

Fig. 93. Drawing of a crystal of penta-erythritol showing the signs and positions of the charges developed on heating. (Reproduced from *Lehrbuch der Kristallphysik*, W. Voigt (Teubner, Leipzig, 1928), by kind permission of the publishers.)

Fig. 94. Drawing of a crystal of tourmaline showing the signs and positions of the charges developed on heating. (Reproduced from *Lehrbuch der Kristallphysik*, W. Voigt (Teubner, Leipzig, 1928), by kind permission of the publishers.)

Tourmaline (complex silicate containing boron and aluminium). This crystal belongs to class 3m of the trigonal system and has a unique trigonal axis.

Boracite (6MgO.MgCl₂.8B₂O₃). This crystal is a rhombic pseudomorph at ordinary temperatures, though at higher

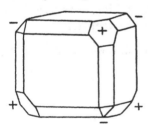

Fig. 95. Drawing of a crystal of boracite showing the signs and positions of the charges developed on heating. (Reproduced from *Crystallography and Practical Crystal Measurement*, A. E. H. Tutton (Macmillan, London, 1922), by kind permission of the publishers.)

temperatures it is cubic with the symmetry $\bar{4}3m$ shown in fig. 95.

7. The pyro-electricity shown by crystals which apparently possess a centre of symmetry.

Certain crystals, notably topaz, axinite and dioptase, show pyro-electricity, although the face development of rare doubly terminated specimens suggests that they have centres of symmetry. This might be taken to invalidate the rule enunciated earlier, paragraph 1, were it not for certain peculiarities of these crystals. Not all topazes are pyro-electric; those from Brazil show a strong effect, while those from Siberia show none. The Brazilian specimens do not have definite electric axes, but the lines of force often concentrate round certain regions of the crystals. Axinite shows a similar pyro-electric character. It is probable that a true single crystal of topaz or axinite would have a polar axis but that natural crystals are more or less twinned. Evidence for this in the case of topaz is afforded by experiments† in which the crystal was powdered, and heated whilst in a strong electric field. The grains orientated themselves and gave a plate which on subsequent heating in the absence of an electric field showed a very strong pyro-electric effect. In conclusion it may be said that untwinned pyro-electric crystals possessing a centre of symmetry have not yet been found, and the rule that pyro-electricity is only shown by crystals lacking a centre of symmetry is as yet without unambiguous exception.

† A. Meissner and R. Bechmann, *Z. techn. Phys.* **9**, 430 (1928).

CHAPTER VIII

ELASTICITY

1. Introduction.

The elastic properties of crystals have been studied for a long time and mathematicians have paid much attention to the problems raised. The same fundamental principles apply to the study of elasticity in crystals as in isotropic bodies with the difference introduced by the symmetry of the crystal. The method of introducing the effects due to symmetry has been completely worked out by various authors, Voigt giving a complete statement in his *Lehrbuch*, which has become classical. In spite of the complete theoretical analysis which has been available for at least twenty-five years, the experimental data that have been collected are very meagre. A few substances which form good crystals have been investigated but no systematic study has been made of crystals as a group. Two new techniques which have greatly widened the scope of inquiry have recently been introduced; one is the application of sound-waves generated by piezo-electric crystals and the other the methods of X-ray crystal structure analysis. The technique of crystal structure analysis makes it possible to study the elastic deformation of the unit cell, and also the relative displacements of the various atoms within the unit cell.

2. Components of stress.

It is necessary to consider the equilibrium between external and internal forces acting on any homogeneous body. It is irrelevant for the purposes of this paragraph whether the body is crystalline or not—the only condition is that from a mechanical point of view it shall be homogeneous. The state of deformation at every point of the body is assumed to be the same, and to obtain the relation between the external forces and the reactions of the deformed body, we shall consider a small tetrahedron bounded by one external and three internal faces parallel to the coordinate axial planes. Suppose, therefore, that O (fig. 96) is a point within the body, and that the axes of reference are OX_1, OX_2, OX_3. OP is normal to the plane ABC which is part of the external surface. A pressure k is supposed to be exerted on ABC. The direction of this pressure will not, in general, coincide with the normal to the

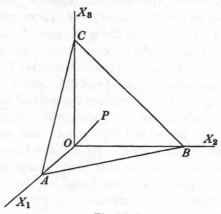

Fig. 96

plane. To balance this pressure forces will act on the faces AOB, BOC and COA and these will be denoted t_{ik}, where t_{ik} is the force acting on the face perpendicular to the ith axis in the direction of the kth axis. For example, t_{33} is the force per unit area acting on face AOB in the direction of X_3, while t_{12} is the force per unit area acting on face BOC in the direction X_2.

In equilibrium the components of k will be equal to the sum of reactions parallel to each component. Thus

$$k_1 \cdot \text{area } ABC = t_{11} \cdot \text{area } BOC + t_{21} \cdot \text{area } COA + t_{31} \cdot \text{area } AOB,$$

that is
$$k_1 = t_{11}l_1 + t_{21}l_2 + t_{31}l_3,$$

where l_1, l_2, l_3 are the direction-cosines of the normal OP.

Similarly, $\qquad k_2 = t_{12}l_1 + t_{22}l_2 + t_{32}l_3,$

or, generally, $\qquad k_i = t_{1i}l_1 + t_{2i}l_2 + t_{3i}l_3.$ $\qquad$ ......(1)

If we consider any unit cube within the body it is in equilibrium as a result of the balanced reactions. Thus, if on the face $ABCD$ (fig. 97) a force t_{21} acts in the direction $+X_1$, then the unit cube under consideration must exert an equal and opposite force on the neighbouring unit cube. The reaction to this is a

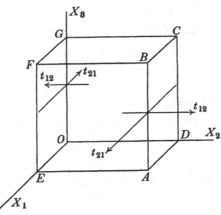

Fig. 97. Diagram showing opposing couples acting on a rectangular block.

force t_{21} acting in the direction $-X_1$ on the face $OEFG$. Similarly equal and oppositely directed forces act on faces $AEFB$ and $DOGC$. The resultant of these two couples must be zero if the cube is not to rotate.

Hence $\qquad\qquad t_{21} = t_{12},$

or, in general, $\qquad\qquad t_{ik} = t_{ki}.$

The quantities t_{ik} are known as the components of stress, and, being defined by a linear relation of the type of equation (1), transform as second order tensors.

The definition of the properties of the components of strain, r_{ik}, has been given in Chapter II, p. 19, where it was shown that

$$r_{ik} = r_{ki}.$$

3. Generalised Hooke's law.†

The simple statement that the stress is proportional to the strain is found to hold experimentally when applied to the components of stress and strain defined above. The general relation involves the fourth order tensor c_{ikpq}, which is the general expression for the *elastic constant*. Thus

$$t_{pq} = c_{ikpq} r_{ik}, \qquad \ldots\ldots(2)$$

or, expanding for a particular case,

$$\begin{aligned}
t_{33} = &\, c_{1133} r_{11} + c_{1233} r_{12} + c_{1333} r_{13} \\
&+ c_{2133} r_{21} + c_{2233} r_{22} + c_{2333} r_{23} \\
&+ c_{3133} r_{31} + c_{3233} r_{32} + c_{3333} r_{33}. \qquad \ldots\ldots(3)
\end{aligned}$$

Each component of stress is linearly dependent upon all the components of strain and conversely. The dependence of strain upon the components of stress is expressed thus:

$$r_{pq} = s_{ikpq} t_{ik}. \qquad \ldots\ldots(4)$$

The quantities s_{ikpq} are known as the *elastic moduli* and can be measured experimentally, whereas, in general, the elastic constants c_{ikpq} cannot. This may be seen from equation (3). If it were possible to apply such a set of components of stress that all strains except one vanished, then the values of c_{ikpq} could be measured, one at a time. For instance, in equation (3) if all r's vanish except r_{33}, then we have

$$t_{33} = c_{3333} r_{33}$$

and

$$c_{3333} = \frac{t_{33}}{r_{33}},$$

but, in general, it is not possible to achieve this. On the other hand, it is easy to apply only one component of stress and from equation (4) obtain the corresponding value of s_{ikpq}.

The interchangeability of the suffixes in r_{ik} and t_{pq} means that it is unnecessary to distinguish i from k or p from q.

† Hooke's law states that, when a body is subjected to stress which does not exceed the elastic limit, the change in shape produced (strain) is proportional to the deforming stress.

It is, therefore, customary, as was mentioned in Chapter VI, to write single numbers instead of double numbers according to the following scheme:

Old numbers	11	22	33	23, 32	31, 13	12, 21
New numbers	1	2	3	4	5	6

Since there are two equivalent stress or strain components when the suffixes are unequal the numbers of components comprised under each of the terms t_4, t_5, t_6 is two, not one. This gives rise to certain differences between the meanings of corresponding moduli in Voigt's *Lehrbuch* and in this book.

4.* The relations between the elastic moduli† and the elastic constants.

We shall now express t_{33} in terms of the new notation, referring to equation (3) for its value in the old. Thus

$$t_3 = c_{13}r_1 + c_{63}r_6 + c_{53}r_5$$
$$+ c_{63}r_6 + c_{23}r_2 + c_{43}r_4$$
$$+ c_{53}r_5 + c_{43}r_4 + c_{33}r_3.$$

Or, collecting terms,

$$t_3 = c_{13}r_1 + c_{23}r_2 + c_{33}r_3 + 2c_{43}r_4 + 2c_{53}r_5 + 2c_{63}r_6.$$

In exactly the same way we may write down the values of $t_1, \ldots, 2t_6$. Thus

$$t_1 = c_{11}r_1 + c_{21}r_2 + c_{31}r_3 + 2c_{41}r_4 + 2c_{51}r_5 + 2c_{61}r_6,$$
$$t_2 = c_{12}r_1 + c_{22}r_2 + c_{32}r_3 + 2c_{42}r_4 + 2c_{52}r_5 + 2c_{62}r_6,$$
$$t_3 = c_{13}r_1 + c_{23}r_2 + c_{33}r_3 + 2c_{43}r_4 + 2c_{53}r_5 + 2c_{63}r_6,$$
$$2t_4 = 2c_{14}r_1 + 2c_{24}r_2 + 2c_{34}r_3 + 4c_{44}r_4 + 4c_{54}r_5 + 4c_{64}r_6,$$
$$2t_5 = 2c_{15}r_1 + 2c_{25}r_2 + 2c_{35}r_3 + 4c_{45}r_4 + 4c_{55}r_5 + 4c_{65}r_6,$$
$$2t_6 = 2c_{16}r_1 + 2c_{26}r_2 + 2c_{36}r_3 + 4c_{46}r_4 + 4c_{56}r_5 + 4c_{66}r_6.$$

† It is important to notice that Young's modulus for a rod directed parallel to X'_3 is equal to $1/s'_{33}$.

These are six simultaneous linear equations in $r_1, ..., r_6$ and we may therefore solve them according to the usual rules. Thus

$$r_1 = \frac{\begin{vmatrix} t_1 & c_{21} & c_{31} & 2c_{41} & 2c_{51} & 2c_{61} \\ t_2 & c_{22} & c_{32} & 2c_{42} & 2c_{52} & 2c_{62} \\ t_3 & c_{23} & c_{33} & 2c_{43} & 2c_{53} & 2c_{63} \\ 2t_4 & 2c_{24} & 2c_{34} & 4c_{44} & 4c_{54} & 4c_{64} \\ 2t_5 & 2c_{25} & 2c_{35} & 4c_{45} & 4c_{55} & 4c_{65} \\ 2t_6 & 2c_{26} & 2c_{36} & 4c_{46} & 4c_{56} & 4c_{66} \end{vmatrix}}{\Delta_c},$$

where Δ_c is the determinant consisting of the c's only. Similarly,

$$r_4 = -\frac{\begin{vmatrix} t_1 & 2c_{51} & 2c_{61} & c_{11} & c_{21} & c_{31} \\ t_2 & 2c_{52} & 2c_{62} & c_{12} & c_{22} & c_{32} \\ t_3 & 2c_{53} & 2c_{63} & c_{13} & c_{23} & c_{33} \\ 2t_4 & 4c_{54} & 4c_{64} & 2c_{14} & 2c_{24} & 2c_{34} \\ 2t_5 & 4c_{55} & 4c_{65} & 2c_{15} & 2c_{25} & 2c_{35} \\ 2t_6 & 4c_{56} & 4c_{66} & 2c_{16} & 2c_{26} & 2c_{36} \end{vmatrix}}{\Delta_c}.$$

If the subdeterminants of Δ_c obtained by omitting the ith row and the kth column from Δ_c be denoted $\delta_{ik(c)}$, then

$$r_i = \frac{\delta_{1i(c)} t_1}{\Delta_c} + \frac{\delta_{2i(c)} t_2}{\Delta_c} + \frac{\delta_{3i(c)} t_3}{\Delta_c} + \frac{\delta_{4i(c)} 2t_4}{\Delta_c}$$
$$+ \frac{\delta_{5i(c)} 2t_5}{\Delta_c} + \frac{\delta_{6i(c)} 2t_6}{\Delta_c}. \quad ......(5)$$

Thus by comparing equations (5) and (4) we see that

$$s_{ki} = \frac{\delta_{ki(c)}}{\Delta_c}.$$

Similarly, if $\delta_{ki(s)}$ and Δ_s denote the corresponding subdeterminants and determinants when the s's replace the c's, it is clear from the similarity of equations (2) and (4) that

$$c_{ki} = \frac{\delta_{ki(s)}}{\Delta_s}.$$

Application to the cubic system.

The full determinant $\varDelta_c$ is as follows:

$$
\begin{vmatrix}
c_{11} & c_{12} & c_{12} & 0 & 0 & 0 \\
c_{12} & c_{11} & c_{12} & 0 & 0 & 0 \\
c_{12} & c_{12} & c_{11} & 0 & 0 & 0 \\
0 & 0 & 0 & 4c_{44} & 0 & 0 \\
0 & 0 & 0 & 0 & 4c_{44} & 0 \\
0 & 0 & 0 & 0 & 0 & 4c_{44}
\end{vmatrix}
$$

Hence

$$
s_{11} = \frac{(4 . c_{44})^3 . \begin{vmatrix} c_{11} & c_{12} \\ c_{12} & c_{11} \end{vmatrix}}{(4 . c_{44})^3 . \begin{vmatrix} c_{11} & c_{12} & c_{12} \\ c_{12} & c_{11} & c_{12} \\ c_{12} & c_{12} & c_{11} \end{vmatrix}}
$$

$$
= \frac{(c_{11}^2 - c_{12}^2)}{c_{11}(c_{11}^2 - c_{12}^2) + c_{12}(c_{12}^2 - c_{11}c_{12}) + c_{12}(c_{12}^2 - c_{11}c_{12})}
$$

$$
= \frac{(c_{11} + c_{12})\,(c_{11} - c_{12})}{(c_{11} - c_{12})\,(c_{11}^2 + c_{11}c_{12} - 2c_{12}^2)}
$$

$$
= \frac{c_{11} + c_{12}}{(c_{11} + 2c_{12})\,(c_{11} - c_{12})}.
$$

Also

$$
s_{44} = \frac{(4c_{44})^2 . \begin{vmatrix} c_{11} & c_{12} & c_{12} \\ c_{12} & c_{11} & c_{12} \\ c_{12} & c_{12} & c_{11} \end{vmatrix}}{(4c_{44})^3 . \begin{vmatrix} c_{11} & c_{12} & c_{12} \\ c_{12} & c_{11} & c_{12} \\ c_{12} & c_{12} & c_{11} \end{vmatrix}} = \frac{1}{4c_{44}}.\dagger
$$

† The elastic constants c_{ik} used here are identical with those employed by Voigt, but the elastic moduli are not. If Voigt's moduli be denoted $s_{ik(V)}$ and those used here $s_{ik(W)}$ then,

$$
s_{ik(V)} = s_{ik(W)} \text{ when } i \text{ and } k \text{ are } 1, 2, \text{ or } 3
$$
$$
= 2s_{ik(W)} \text{ when either } i \text{ or } k \text{ is } 4, 5, \text{ or } 6
$$
$$
= 4s_{ik(W)} \text{ when both } i \text{ and } k \text{ are } 4, 5, \text{ or } 6.
$$

5.* The equalities $c_{ik} = c_{ki}$ and $s_{ik} = s_{ki}$.

If the deformation at any point is given by the quantities $(r_1, \ldots, r_6)$ and these are changed to $(r_1 + dr_1, \ldots, r_6 + dr_6)$ by an increase in stress, the potential energy, Φ, associated with the deformation is increased by an amount $\delta\Phi$ which is given by

$$\delta\Phi = -(t_1 dr_1 + \ldots + 2t_6 dr_6).$$

Since Φ depends only on the six r's and they are all independent,

$$d\Phi = \frac{\partial\Phi}{\partial r_1} dr_1 + \ldots + \frac{\partial\Phi}{\partial r_6} dr_6.$$

Hence,
$$t_1 = -\frac{\partial\Phi}{\partial r_1},$$

$$2t_6 = -\frac{\partial\Phi}{\partial r_6}.$$

Now
$$t_1 = c_{11} r_1 + \ldots + 2c_{61} r_6,$$
$$\vdots$$
$$t_6 = c_{16} r_1 + \ldots + 2c_{66} r_6,$$

and, therefore,
$$\frac{\partial t_1}{\partial r_2} = c_{21} = -\frac{\partial^2\Phi}{\partial r_1 \partial r_2};$$

similarly,
$$\frac{\partial t_2}{\partial r_1} = c_{12} = -\frac{\partial^2\Phi}{\partial r_2 \partial r_1}$$

and
$$c_{12} = c_{21}.$$

In the same way it may be shown that in general

$$c_{ik} = c_{ki}.$$

If in the expression derived in the previous paragraph, namely, $\delta_{ik(c)}/\Delta_{(c)}$, k be interchanged for i no change will result, since $c_{ik} = c_{ki}$. Hence

$$s_{ik} = s_{ki}.$$

6.* Application of stress to a parallelepiped the faces of which are parallel to the planes containing the axes of coordinates.

If compression is applied along the X_1 axis but not in any other direction, then all components of stress except t_1 are zero.

From equation (4) we obtain

$$r_1 = s_{11}t_1, \qquad r_4 = s_{14}t_1,$$
$$r_2 = s_{12}t_1, \qquad r_5 = s_{15}t_1,$$
$$r_3 = s_{13}t_1, \qquad r_6 = s_{16}t_1.$$

Thus, for a compression along X_1,

s_{11} is the modulus giving the contraction in the X_1 direction,

s_{12} ,,   ,,   ,,   ,, X_2 ,,

s_{13} ,,   ,,   ,,   ,, X_3 ,,

s_{14} is the modulus giving the change of angle between the X_2 and X_3 axes,

s_{15} is the modulus giving the change of angle between the X_3 and X_1 axes,

s_{16} is the modulus giving the change of angle between the X_1 and X_2 axes.

The volume expansion due to this force in the X_1 direction may be found as follows:

r_1, r_2, r_3 are the changes in length per unit length parallel to the three edges of the parallelepiped and hence the new volume of a unit cube is $1 + \delta$, where

$$\delta = r_1 + r_2 + r_3$$
$$= (s_{11} + s_{12} + s_{13})\, t_1.$$

7.* Uniform hydrostatic pressure acting on a rectangular parallelepiped—two sets of principal elasticity axes.

In this case $t_1 = t_2 = t_3 = \pi$

and $t_4 = t_5 = t_6 = 0.$

Whence
$$r_1 = s_{11}t_1 + s_{21}t_2 + s_{31}t_3$$
$$= (s_{11} + s_{21} + s_{31})\,\pi,$$
$$r_2 = (s_{12} + s_{22} + s_{32})\,\pi,$$
$$\vdots$$
$$r_6 = (s_{16} + s_{26} + s_{36})\,\pi.$$

These equations contain no term involving the coordinates of a point in the elastic body and are therefore valid for all points of it. Thus the components of strain are the same at every point of the body and the deformation is homogeneous (see Chapter II).

From the theory of homogeneous deformation it is always found possible to choose such a set of mutually perpendicular axes that a deformation under hydrostatic pressure does not change the angles between them.

If such axes have been chosen, then
$$r_4 = r_5 = r_6 = 0,$$
so that
$$s_{14} + s_{24} + s_{34} = s_{15} + s_{25} + s_{35} = s_{16} + s_{26} + s_{36} = 0.$$

These relations involving the elastic moduli define the axes which remain mutually perpendicular on being subjected to a uniform hydrostatic pressure.

We can similarly define another set of mutually perpendicular axes which remain unchanged when a uniform dilatation or contraction is caused by the components of the stresses involving simple pressure.

If we put
$$r_1 = r_2 = r_3 = p,$$
$$r_4 = r_5 = r_6 = 0,$$
then
$$t_1 = c_{11}r_1 + c_{21}r_2 + c_{31}r_3 = p(c_{11} + c_{21} + c_{31}),$$
$$t_2 = c_{12}r_1 + c_{22}r_2 + c_{32}r_3 = p(c_{12} + c_{22} + c_{32}),$$
$$\vdots$$
$$t_6 = p(c_{16} + c_{26} + c_{36}).$$

If the shear components of stress are zero, then
$$c_{14} + c_{24} + c_{34} = c_{15} + c_{25} + c_{35} = c_{16} + c_{26} + c_{36} = 0.$$

These equations define another set of axes which are not necessarily the same as those fixed by the corresponding equations involving s's. In crystal systems of high symmetry the two sets of axes coincide, but in the monoclinic system only one axis need be common to them both, and in the triclinic system, none at all.

8.* The elastic constants and moduli which occur in the various crystal systems.

A 4th order tensor has 81 components, but many of these may be equal. The elasticity tensor, for instance, reduces to 21 different components in the general case, (1) because $t_{ik} = t_{ki}$ and $r_{ik} = r_{ki}$, making possible the substitution of one subscript, ranging from 1 to 6, and thereby giving only 6^2 different s_{ik}'s; (2) because $s_{ik} = s_{ki}$, reducing the 36 components to 21.

Triclinic system.

There is no limitation imposed on the constants by the symmetry.

Monoclinic system.

We take the diad axis to be X_3. The rotation characteristic of a diad axis may be expressed by transforming $+X_1$ into $-X_1$, $+X_2$ into $-X_2$ and leaving X_3 unchanged. The direction-cosines of the new axes relative to the old are given by the scheme

$$
\begin{array}{cccc}
 & X_1' & X_2' & X_3' \\
X_1 & -1 & 0 & 0 \\
X_2 & 0 & -1 & 0 \\
X_3 & 0 & 0 & 1
\end{array}
$$

We therefore obtain a transformation†

$$s_{1111}' = \alpha_{11}\alpha_{11}\alpha_{11}\alpha_{11}s_{1111} = s_{1111},$$

$$s_{1122}' = \alpha_{11}\alpha_{11}\alpha_{22}\alpha_{22}s_{1122} = s_{1122}, \text{ etc.},$$

$$s_{1113}' = \alpha_{11}\alpha_{11}\alpha_{11}\alpha_{33}s_{1113} = -s_{1113} = 0,$$

† To avoid confusion with the elastic constants direction-cosines are here denoted α_{ik}.

and similarly for any s containing one 3 or three 3's. Using the abbreviated nomenclature, the moduli which do not vanish in this system are therefore

$$
\begin{array}{cccccc}
s_{11} & s_{12} & s_{13} & 0 & 0 & s_{16} \\
 & s_{22} & s_{23} & 0 & 0 & s_{26} \\
 & & s_{33} & 0 & 0 & s_{36} \\
 & & & s_{44} & s_{45} & 0 \\
 & & & & s_{55} & 0 \\
 & & & & & s_{66}
\end{array}
$$

Rhombic system.

The existence of three mutually perpendicular diad axes necessitates that all moduli having in the full nomenclature only one or three of any of the numbers 1, 2, 3 must vanish. Thus the non-vanishing moduli are

$$
\begin{array}{cccccc}
s_{11} & s_{12} & s_{13} & 0 & 0 & 0 \\
 & s_{22} & s_{23} & 0 & 0 & 0 \\
 & & s_{33} & 0 & 0 & 0 \\
 & & & s_{44} & 0 & 0 \\
 & & & & s_{55} & 0 \\
 & & & & & s_{66}
\end{array}
$$

Tetragonal system.

For a rotation of 90° the transformation scheme is

$$
\begin{array}{cccc}
 & X_1' & X_2' & X_3' \\
X_1 & 0 & -1 & 0 \\
X_2 & 1 & 0 & 0 \\
X_3 & 0 & 0 & 1
\end{array}
$$

and for a rotation of 180° about the X_3 axis we have the same limitations as apply to the monoclinic system. The requirements of a 90° rotation will therefore be applied to each of the non-zero constants of the monoclinic system.

$$s'_{1111} = \alpha_{21}\alpha_{21}\alpha_{21}\alpha_{21}s_{2222} = s_{2222} = s_{1111}, \quad \text{i.e. } s_{11} = s_{22},$$

$$s'_{1122} = \alpha_{21}\alpha_{21}\alpha_{12}\alpha_{12}s_{2211} = s_{2211} = s_{1122}, \quad \text{i.e. } s_{12} \neq 0,$$

$$s'_{1133} = \alpha_{21}\alpha_{21}\alpha_{33}\alpha_{33}s_{2233} = s_{2233} = s_{1133}, \quad \text{i.e. } s_{13} = s_{23},$$

$$s'_{2323} = \alpha_{12}\alpha_{33}\alpha_{12}\alpha_{33}s_{1313} = s_{1313} = s_{2323}, \quad \text{i.e. } s_{44} = s_{55},$$

$$s'_{1112} = \alpha_{21}\alpha_{21}\alpha_{21}\alpha_{12}s_{2221} = -s_{2221} = s_{1112}, \quad \text{i.e. } s_{16} = -s_{26},$$

$$s'_{3312} = \alpha_{33}\alpha_{33}\alpha_{21}\alpha_{12}s_{3321} = -s_{3321} = -s_{3312} = s_{3312} = 0,$$

$$\text{i.e. } s_{36} = 0,$$

$$s'_{2313} = \alpha_{12}\alpha_{33}\alpha_{21}\alpha_{33}s_{1323} = -s_{1323} = -s_{2313} = 0, \quad \text{i.e. } s_{45} = 0.$$

Similarly $\quad\quad\quad\quad\quad s_{46} = s_{56} = 0.$

After examining all the moduli in this way the scheme is found to be

$$\begin{matrix} s_{11} & s_{12} & s_{13} & 0 & 0 & s_{16} \\ & s_{11} & s_{13} & 0 & 0 & -s_{16} \\ & & s_{33} & 0 & 0 & 0 \\ & & & s_{44} & 0 & 0 \\ & & & & s_{44} & 0 \\ & & & & & s_{66} \end{matrix}$$

When horizontal diad axes are present $s_{16} = 0$, because it contains three 1's.

Cubic system.

There must be in this system the possibility of cyclical interchange of the three axes without change of physical properties. Hence we must be able to change 1 into 2, 2 into 3, 3 into 1, without change of modulus. Thus the cubic scheme becomes

$$\begin{matrix} s_{11} & s_{12} & s_{12} & 0 & 0 & 0 \\ & s_{11} & s_{12} & 0 & 0 & 0 \\ & & s_{11} & 0 & 0 & 0 \\ & & & s_{44} & 0 & 0 \\ & & & & s_{44} & 0 \\ & & & & & s_{44} \end{matrix}$$

Trigonal and hexagonal systems.

The derivation of the scheme of moduli for these systems is rather long, and will not be given in detail. The direction-cosines of the transformed axes are expressed in terms of the sine and cosine of the angle of rotation about X_3 (see Chapter I, p. 11). The expression for s'_{1111} contains only terms in s_{1111}, s_{1112}, s_{1122}, s_{1212}, s_{1222}, s_{2222} (remembering the equivalence of s_{ikpq}, s_{ikqp}, s_{kipq}, s_{kiqp}, s_{pqik}, s_{qpik}, s_{pqki}, s_{qpki}). By equating s'_{1111} and s_{1111} we obtain one equation connecting the six s's above. Similar transformations of s'_{1112}, s'_{1122}, etc. yield six such equations. The solution of these equations leads to the results

$$s_{1111} = s_{2222} \qquad \text{or} \qquad s_{11} = s_{22},$$
$$s_{1112} = s_{1222} = 0 \qquad \text{or} \qquad s_{16} = s_{26} = 0,$$
$$s_{1212} = \frac{s_{1111} - s_{1122}}{2} \qquad \text{or} \qquad s_{66} = \frac{s_{11} - s_{12}}{2}.$$

By transforming s'_{1113} and equating the result to s_{1113} we obtain an equation connecting s_{1113}, s_{1123}, s_{1213}, s_{1223}, s_{2213}, s_{2223}. The transformations corresponding to each of these moduli provide six equations, which show that for a rotation of 120° about X_3,

$$s_{1123} = -s_{2223} = s_{1312} \quad \text{or} \quad s_{14} = -s_{24} = s_{56},$$
$$s_{1113} = -s_{2213} = -s_{2312} \quad \text{or} \quad s_{15} = -s_{25} = -s_{46}.$$

When the rotation is 60° we have in addition the relations

$$s_{14} = s_{15} = 0.$$

The transformations of s'_{1133} and s'_{1233} lead to two equations, from which it follows that

$$s_{1133} = s_{2233} \qquad \text{or} \qquad s_{13} = s_{23},$$
$$s_{1233} = 0 \qquad \text{or} \qquad s_{36} = 0.$$

The transformations of s'_{1313} and s'_{1323} lead to two equations, which give the results

$$s_{1313} = s_{2323} \qquad \text{or} \qquad s_{44} = s_{55},$$
$$s_{1323} = 0 \qquad \text{or} \qquad s_{45} = 0.$$

Finally, the transformations of s'_{1333} and s'_{2333} enable the following results to be derived:

$$s_{1333} = s_{2333} = 0 \quad \text{or} \quad s_{35} = s_{34} = 0.$$

The schemes of moduli for the two systems are therefore as follows:

Trigonal system (classes 3, $\bar{3}$) (C_3, C_{3i}).

$$
\begin{array}{cccccc}
s_{11} & s_{12} & s_{13} & s_{14} & -s_{25} & 0 \\
 & s_{11} & s_{13} & -s_{14} & s_{25} & 0 \\
 & & s_{33} & 0 & 0 & 0 \\
 & & & s_{44} & 0 & s_{25} \\
 & & & & s_{44} & s_{14} \\
 & & & & & \tfrac{1}{2}(s_{11}-s_{12})
\end{array}
$$

In classes 32, $\bar{3}m$ and $3m$ $(D_3, D_{3d}$ and $C_{3v})$, where a horizontal axis coincides† with X_1 all those s's possessing only one 1 or three 1's in the full notation must vanish—in which case $s_{25} = 0$.

$$
\begin{array}{cccccc}
s_{11} & s_{12} & s_{13} & s_{14} & 0 & 0 \\
 & s_{11} & s_{13} & -s_{14} & 0 & 0 \\
 & & s_{33} & 0 & 0 & 0 \\
 & & & s_{44} & 0 & 0 \\
 & & & & s_{44} & s_{14} \\
 & & & & & \tfrac{1}{2}(s_{11}-s_{12})
\end{array}
$$

Hexagonal system (all classes).

$$
\begin{array}{cccccc}
s_{11} & s_{12} & s_{13} & 0 & 0 & 0 \\
 & s_{11} & s_{13} & 0 & 0 & 0 \\
 & & s_{33} & 0 & 0 & 0 \\
 & & & s_{44} & 0 & 0 \\
 & & & & s_{44} & 0 \\
 & & & & & \tfrac{1}{2}(s_{11}-s_{12})
\end{array}
$$

† A plane of symmetry perpendicular to X_1 is equivalent to a diad axis parallel to X_1

9. Representation surfaces for elastic properties.

The relation between stresses and the strains that they produce may be represented by a number of surfaces, of which two are important. In one each radius vector is proportional to the bending modulus s'_{33} in that direction, and in the other each radius vector is proportional to the rigidity modulus, $2(s'_{44} + s'_{55})$,† of a cylinder having its axis in that direction. No general description of these surfaces is possible because they vary from substance to substance within the same crystal class, depending for their shape on the relative values of the moduli. The following equations and figures represent the properties of four important crystals belonging to different crystal systems.

9.1. Rochelle salt. $NaKC_4H_4O_6 . 4H_2O$ (Rhombic).

The appropriate equations for the bending and rigidity moduli in a direction defined by the direction-cosines α_{13}, α_{23}, α_{33} may be obtained by writing out the full expressions for s'_{3333}, s'_{2323} and s'_{1313} in terms of s_{iklm}. The reduced expressions are as follows:

$$s'_{33}\ddagger = \alpha_{13}^4 s_{11} + \alpha_{23}^4 s_{22} + \alpha_{33}^4 s_{33}$$
$$+ (4s_{44} + 2s_{23})\, \alpha_{23}^2 \alpha_{33}^2 + (4s_{55} + 2s_{31})\, \alpha_{33}^2 \alpha_{13}^2$$
$$+ (4s_{66} + 2s_{12})\, \alpha_{13}^2 \alpha_{23}^2,$$

$$2(s'_{44} + s'_{55}) = 2[\alpha_{13}^2(s_{55} + s_{66}) + \alpha_{23}^2(s_{66} + s_{44}) + \alpha_{33}^2(s_{44} + s_{55})$$
$$+ \alpha_{23}^2 \alpha_{33}^2(s_{22} + s_{33} - 4s_{44} - 2s_{23})$$
$$+ \alpha_{33}^2 \alpha_{13}^2(s_{33} + s_{11} - 4s_{55} - 2s_{31})$$
$$+ \alpha_{13}^2 \alpha_{23}^2(s_{11} + s_{22} - 4s_{66} - 2s_{12})].$$

† A formal proof that s'_{3333} is the appropriate modulus for the bending of a long thin plate cut parallel to the X'_3 axis is too long to give here. It may be found in Voigt's *Lehrbuch*, p. 617. The same applies to the moduli $4s'_{2323}$ and $2(s'_{2323} + s'_{3131})$ which apply respectively to the twisting of a plate and a circular cylinder elongated parallel to X'_3.

‡ The factors 2 and 4 arise from the number of ways in which the sets of four suffixes may be written, e.g. 2323, 2332, 3223 and 3232. The terms with these suffixes are, of course, all equal.

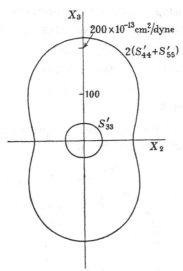

Fig. 98. Sections parallel to (100) of the elasticity surfaces corresponding to bending a plate, s'_{33}, and twisting a cylinder $2(s'_{44}+s'_{55})$ of Rochelle salt.

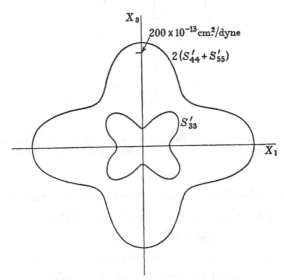

Fig. 99. Sections parallel to (010) of the elasticity surfaces corresponding to bending a plate, s'_{33}, and twisting a cylinder $2(s'_{44}+s'_{55})$ of Rochelle salt.

When the following values for the moduli are inserted:

$$s_{11} = 55\cdot7, \qquad s_{33} = 37\cdot3 \times 10^{-13} \text{ cm.}^2/\text{dyne},$$
$$s_{12} = -8\cdot7, \qquad 4s_{44} = 87\cdot4,$$
$$s_{22} = 38\cdot4, \qquad 4s_{55} = 359\cdot6,$$
$$s_{23} = -5\cdot0, \qquad 4s_{66} = 118\cdot3,$$
$$s_{31} = -34\cdot2,$$

figures 98, 99 and 100 represent the principal sections of the two surfaces which are obtained.

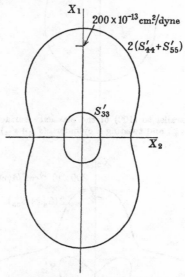

Fig. 100. Sections parallel to (001) of the elasticity surfaces corresponding to bending a plate, s'_{33}, and twisting a cylinder $2(s'_{44} + s'_{55})$ of Rochelle salt.

9.2. Quartz. SiO_2 (Trigonal).

The formulae for the bending and rigidity moduli are

$$s'_{33} = s_{11}(1-\alpha_{33}^2)^2 + s_{33}\alpha_{33}^4 + (4s_{44}+2s_{13})\alpha_{33}^2(1-\alpha_{33}^2)$$
$$+ 4s_{14}\alpha_{23}\alpha_{33}(3\alpha_{13}^2-\alpha_{23}^2),$$

$$2(s'_{44}+s'_{55}) = 2\{s_{44}+\tfrac{1}{2}(s_{11}-s_{12}) + [s_{44}-\tfrac{1}{2}(s_{11}-s_{12})]\alpha_{33}^2$$
$$+ (s_{11}+s_{33}-4s_{44}-2s_{13})\alpha_{33}^2(1-\alpha_{33}^2)$$
$$- 4s_{14}\alpha_{23}\alpha_{33}(3\alpha_{13}^2-\alpha_{23}^2)\}.$$

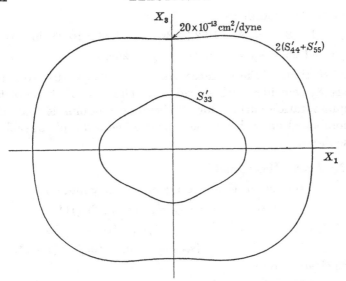

Fig. 101. Sections parallel to the diad and triad axes of the elasticity surfaces for bending a plate, s'_{33}, and twisting a cylinder $2(s'_{44} + s'_{55})$ of quartz.

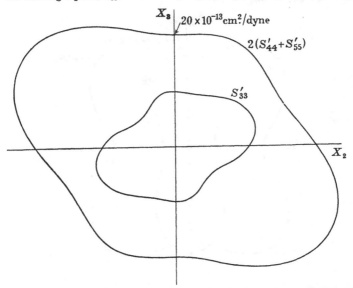

Fig. 102. Sections perpendicular to the diad axis of the elasticity surfaces of quartz corresponding to bending a plate, s'_{33}, and twisting a cylinder $2(s'_{44} + s'_{55})$.

The elastic moduli are

$$s_{11} = 12\cdot98, \quad s_{13} = -1\cdot52, \quad 2s_{14} = -4\cdot31 \times 10^{-13} \text{ cm.}^2/\text{dyne},$$
$$s_{12} = -1\cdot66, \quad s_{33} = 9\cdot90, \quad 4s_{44} = 20\cdot04,$$

and sections of the surfaces are given in figs. 101, 102. The axis X_1 coincides with a diad axis. The general shape of the representation surface for the bending modulus is that of a rhombohedron with rounded corners and a dimple in each of the faces.

9.3. Zinc. (Hexagonal.)

The formulae for the bending and rigidity moduli are

$$s_{33}' = s_{11}(1-\alpha_{33}^2)^2 + s_{33}\alpha_{33}^4 + (4s_{44}+2s_{13})(1-\alpha_{33}^2)\alpha_{33}^2,$$
$$2(s_{44}'+s_{55}') = 2\{s_{44}+\tfrac12(s_{11}-s_{12}) + [s_{44}-\tfrac12(s_{11}-s_{12})]\alpha_{33}^2$$
$$+ (s_{11}+s_{33}-4s_{44}-2s_{13})\alpha_{33}^2(1-\alpha_{33}^2)\}.$$

The elastic moduli are

$$s_{11} = 7\cdot70, \quad s_{12} = 0\cdot83, \quad s_{13} = -6\cdot93 \times 10^{-13} \text{ cm.}^2/\text{dyne},$$
$$s_{33} = 27\cdot66, \quad 4s_{44} = 24\cdot40,$$

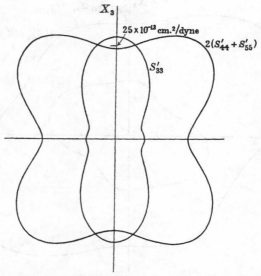

Fig. 103. Sections parallel to the hexad axis of the elasticity surfaces of zinc corresponding to bending a plate, s'_{33}, and twisting a cylinder $2(s'_{44}+s'_{55})$.

and sections of the corresponding surfaces are represented in fig. 103. The appearance of only one direction-cosine, α_{33}, in the formulae implies that both surfaces are surfaces of revolution.

9.4. Rock-salt. (Cubic.)

The formulae for the bending and rigidity moduli are

$$s'_{33} = \tfrac{1}{2}(4s_{44} + 2s_{12}) + (s_{11} - s_{12} - 2s_{44})(\alpha_{13}^4 + \alpha_{23}^4 + \alpha_{33}^4),$$

$$2(s'_{44} + s'_{55}) = 2\{2s_{44} + [2(s_{11} - s_{12}) - 4s_{44}]$$
$$\times (\alpha_{23}^2\alpha_{33}^2 + \alpha_{33}^2\alpha_{13}^2 + \alpha_{13}^2\alpha_{23}^2)\}.$$

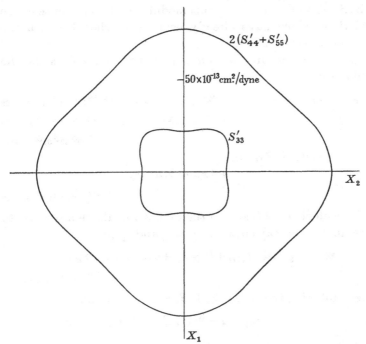

Fig. 104. Sections parallel to (001) of the elasticity surfaces of rock-salt corresponding to bending a plate, s'_{33}, and twisting a cylinder $2(s'_{44} + s'_{55})$.

The elastic moduli are

$$s_{11} = 22\cdot08, \quad s_{12} = -4\cdot49, \quad 4s_{44} = 78\cdot26 \times 10^{-13}\ \text{cm.}^2/\text{dyne},$$

and sections of the corresponding surfaces are represented in fig. 104. The octahedral planes cut both surfaces in circular sections and the general character is that of cubes with rounded corners and with bumps or dimples at the centre of each face.

10.* The determination of all the elastic moduli in the cubic system.

10.1.* The variation of the bending modulus with direction.

The value of s'_{3333} in any arbitrary direction making direction-cosines α_{13}, α_{23}, α_{33} with the crystallographic axes is obtained by transforming this modulus and making use of the relations which have already been established between the moduli in this system.

The 81 terms in the complete expression reduce in the cubic case to

$$
\begin{aligned}
s'_{3333} = {}& \alpha_{13}\alpha_{13}\alpha_{13}\alpha_{13}s_{1111} + 2\alpha_{13}\alpha_{13}\alpha_{23}\alpha_{23}s_{1122} + 2\alpha_{13}\alpha_{13}\alpha_{33}\alpha_{33}s_{1122} \\
& + \alpha_{23}\alpha_{23}\alpha_{23}\alpha_{23}s_{1111} + 2\alpha_{23}\alpha_{23}\alpha_{33}\alpha_{33}s_{1122} \\
& + \alpha_{33}\alpha_{33}\alpha_{33}\alpha_{33}s_{1111} \\
& + 4\alpha_{23}\alpha_{33}\alpha_{23}\alpha_{33}s_{2323} \\
& + 4\alpha_{33}\alpha_{13}\alpha_{33}\alpha_{13}s_{2323} \\
& + 4\alpha_{13}\alpha_{23}\alpha_{13}\alpha_{23}s_{2323}
\end{aligned}
$$

(the coefficients 2 and 4 in the expression above are equal to the number of ways of writing s_{1122} and s_{2323})

$$
\begin{aligned}
= {}& (\alpha_{13}^4 + \alpha_{23}^4 + \alpha_{33}^4)\,s_{1111} + (\alpha_{13}^2\alpha_{23}^2 + \alpha_{23}^2\alpha_{33}^2 + \alpha_{33}^2\alpha_{13}^2) \\
& \times (2s_{1122} + 4s_{2323}).
\end{aligned}
$$

Remembering that $\alpha_{13}^2 + \alpha_{23}^2 + \alpha_{33}^2 = 1$, we see that

$$
\alpha_{13}^4 + \alpha_{23}^4 + \alpha_{33}^4 = 1 - 2(\alpha_{13}^2\alpha_{23}^2 + \alpha_{23}^2\alpha_{33}^2 + \alpha_{33}^2\alpha_{13}^2).
$$

Hence

$$
s'_{3333} = s_{1111} - 2(s_{1111} - s_{1122} - 2s_{2323})\,(\alpha_{13}^2\alpha_{23}^2 + \alpha_{23}^2\alpha_{33}^2 + \alpha_{33}^2\alpha_{13}^2).
$$

From the form of this expression it follows that s_{1111} and $(2s_{1122} + 4s_{2323})$ can be found from measurements of s'_{3333} in various directions. These measurements cannot, however, give the values of $2s_{1122}$ and $4s_{2323}$ separately.

The general expression for the bending modulus in any direction reduces in special cases to the following:

(a) for a plate parallel to a crystallographic axis, say X_1,

$$\alpha_{13} = 1, \quad \alpha_{23} = \alpha_{33} = 0,$$
$$s'_{33} = s_{11};$$

(b) for a plate parallel to an axis making an angle of 45° with two crystallographic axes

$$\alpha_{13} = \alpha_{23} = 1/\sqrt{2}, \quad \alpha_{33} = 0,$$
$$s'_{33} = \tfrac{1}{2}s_{11} + \tfrac{1}{4}(2s_{12} + 4s_{44})$$
$$= \tfrac{1}{2}(s_{11} + s_{12} + 2s_{44});$$

(c) for a plate parallel to a triad axis

$$\alpha_{13} = \alpha_{23} = \alpha_{33} = 1/\sqrt{3},$$
$$s'_{33} = \tfrac{1}{3}(s_{11}) + \tfrac{1}{3}(2s_{12} + 4s_{44})$$
$$= \tfrac{1}{3}(s_{11} + 2s_{12} + 4s_{44}).$$

10.2.* The variation of the rigidity modulus with direction.

Let a plate having its length parallel to the X'_3 axis and its breadth parallel to the X'_2 axis be twisted about its long axis, then the relevant modulus is $4s'_{2323}$ (see footnote, p. 246). Bearing in mind which of the moduli are zero, we may write

$$s'_{2323} = \alpha_{12}\alpha_{13}\alpha_{12}\alpha_{13}s_{1111} + 2\alpha_{12}\alpha_{13}\alpha_{22}\alpha_{23}s_{1122} + 2\alpha_{12}\alpha_{13}\alpha_{32}\alpha_{33}s_{1133}$$
$$+ \; \alpha_{22}\alpha_{23}\alpha_{22}\alpha_{23}s_{2222} + 2\alpha_{22}\alpha_{23}\alpha_{32}\alpha_{33}s_{2233}$$
$$+ \; \alpha_{32}\alpha_{33}\alpha_{32}\alpha_{33}s_{3333}$$
$$+ \alpha_{22}\alpha_{33}\alpha_{22}\alpha_{33}s_{2323} + \alpha_{22}\alpha_{33}\alpha_{32}\alpha_{23}s_{2332}$$
$$+ \alpha_{32}\alpha_{23}\alpha_{22}\alpha_{33}s_{3223} + \alpha_{32}\alpha_{23}\alpha_{32}\alpha_{23}s_{3232}$$
$$+ \alpha_{32}\alpha_{13}\alpha_{32}\alpha_{13}s_{3131} + \alpha_{32}\alpha_{13}\alpha_{12}\alpha_{33}s_{3113}$$
$$+ \alpha_{12}\alpha_{33}\alpha_{32}\alpha_{13}s_{1331} + \alpha_{12}\alpha_{33}\alpha_{12}\alpha_{33}s_{1313}$$
$$+ \alpha_{12}\alpha_{23}\alpha_{12}\alpha_{23}s_{1212} + \alpha_{12}\alpha_{23}\alpha_{22}\alpha_{13}s_{1221}$$
$$+ \alpha_{22}\alpha_{13}\alpha_{12}\alpha_{23}s_{2112} + \alpha_{22}\alpha_{13}\alpha_{22}\alpha_{13}s_{2121}.$$

Inserting the relations

$$s_{11} = s_{22} = s_{33}, \quad s_{12} = s_{13} = s_{23}, \quad s_{44} = s_{55} = s_{66},$$

we have

$$s'_{44} = (\alpha_{12}^2\alpha_{13}^2 + \alpha_{22}^2\alpha_{23}^2 + \alpha_{32}^2\alpha_{33}^2)\,s_{11}$$
$$+ 2(\alpha_{12}\alpha_{13}\alpha_{22}\alpha_{23} + \alpha_{12}\alpha_{13}\alpha_{32}\alpha_{33} + \alpha_{22}\alpha_{23}\alpha_{32}\alpha_{33})\,s_{12}$$
$$+ \{\alpha_{33}^2\alpha_{22}^2 + \alpha_{33}^2\alpha_{12}^2 + \alpha_{32}^2\alpha_{23}^2 + \alpha_{32}^2\alpha_{13}^2 + \alpha_{12}^2\alpha_{23}^2 + \alpha_{13}^2\alpha_{22}^2$$
$$+ 2(\alpha_{12}\alpha_{13}\alpha_{22}\alpha_{23} + \alpha_{12}\alpha_{13}\alpha_{32}\alpha_{33} + \alpha_{22}\alpha_{23}\alpha_{32}\alpha_{33})\}\,s_{44}.$$

Now

$$(\alpha_{12}\alpha_{13} + \alpha_{22}\alpha_{23} + \alpha_{32}\alpha_{33})^2 = \alpha_{12}^2\alpha_{13}^2 + \alpha_{22}^2\alpha_{23}^2 + \alpha_{32}^2\alpha_{33}^2$$
$$+ 2(\alpha_{12}\alpha_{13}\alpha_{22}\alpha_{23} + \alpha_{22}\alpha_{23}\alpha_{32}\alpha_{33} + \alpha_{32}\alpha_{33}\alpha_{12}\alpha_{13}) = 0$$

and
$$\alpha_{12}\alpha_{13}\alpha_{22}\alpha_{23} + \alpha_{22}\alpha_{23}\alpha_{32}\alpha_{33} + \alpha_{32}\alpha_{33}\alpha_{12}\alpha_{13}$$
$$= -\tfrac{1}{2}(\alpha_{12}^2\alpha_{13}^2 + \alpha_{22}^2\alpha_{23}^2 + \alpha_{32}^2\alpha_{33}^2).$$

Also
$$(\alpha_{12}^2 + \alpha_{22}^2 + \alpha_{32}^2)\,(\alpha_{13}^2 + \alpha_{23}^2 + \alpha_{33}^2) = \alpha_{12}^2\alpha_{13}^2 + \alpha_{22}^2\alpha_{13}^2 + \alpha_{32}^2\alpha_{13}^2$$
$$+ \alpha_{12}^2\alpha_{23}^2 + \alpha_{22}^2\alpha_{23}^2 + \alpha_{32}^2\alpha_{23}^2 + \alpha_{12}^2\alpha_{33}^2 + \alpha_{22}^2\alpha_{33}^2 + \alpha_{32}^2\alpha_{33}^2 = 1.$$

Hence
$$\alpha_{33}^2\alpha_{22}^2 + \alpha_{33}^2\alpha_{12}^2 + \alpha_{32}^2\alpha_{23}^2 + \alpha_{32}^2\alpha_{13}^2 + \alpha_{12}^2\alpha_{23}^2 + \alpha_{13}^2\alpha_{22}^2$$
$$= 1 - (\alpha_{12}^2\alpha_{13}^2 + \alpha_{22}^2\alpha_{23}^2 + \alpha_{32}^2\alpha_{33}^2).$$

Thus
$$4s'_{44} = 4(\alpha_{12}^2\alpha_{13}^2 + \alpha_{22}^2\alpha_{23}^2 + \alpha_{32}^2\alpha_{33}^2)\,(s_{11} - s_{12} - 2s_{44}) + 4s_{44}.$$

Special cases.

(a) Twisting about a crystallographic axis, say X_3, when

$$\alpha_{33} = 1 \text{ and all other } \alpha\text{'s are zero,}$$

$$s'_{44} = s_{44}.$$

(b) Twisting about an axis at $45°$ to two crystallographic axes, so that

$$\alpha_{13} = \alpha_{23} = 1/\sqrt{2}, \quad \alpha_{33} = 0$$

when
$$s'_{44} = \tfrac{1}{2}(\alpha_{12}^2 + \alpha_{22}^2)\,(s_{11} - s_{12} - 2s_{44}) + s_{44}$$
$$= \tfrac{1}{2}(s_{11} - s_{12}).$$

(c) Twisting about a trigonal axis,

$$\alpha_{13} = \alpha_{23} = \alpha_{33} = 1/\sqrt{3}$$

and
$$s'_{44} = \tfrac{1}{3}(s_{11} - s_{12} - 2s_{44}) + s_{44}$$
$$= \tfrac{1}{3}(s_{11} - s_{12}) + \tfrac{1}{3}s_{44}.$$

The simplest way of obtaining all three moduli is therefore by combining measurements on bending and twisting. Bending a plate the length of which is parallel to a crystallographic axis gives s_{11}, while twisting the same plate gives s_{44}. A second plate cut parallel to a line at 45° to two of the crystallographic axes or parallel to a triad axis gives a bending or twisting modulus which contains all these moduli and so enables s_{12} to be found.

10.3.* Compressibility modulus.

The change in length per unit length in the X_1 direction is given by r_{11} (see p. 19) and the value of this quantity is related to the components of stress by the relation

$$r_{11} = s_{1111}t_{11} + 2s_{1211}t_{12} + 2s_{1311}t_{13}$$
$$+ s_{2211}t_{22} + 2s_{2311}t_{23}$$
$$+ s_{3311}t_{33}.$$

Now for a cubic crystal

$$s_{1211} = s_{1311} = s_{2311} = 0$$

and

$$s_{2211} = s_{3311} = s_{1122};$$

hence

$$r_{11} = s_{1111}t_{11} + s_{1122}(t_{22} + t_{33}).$$

Under uniform hydrostatic pressure $t_{11} = t_{22} = t_{33}$, and therefore, using the abbreviated notation,

$$r_1 = (s_{11} + 2s_{12})t_1.$$

The cubical compression per unit volume is $3r_1/t_1$, and the modulus S is given by

$$S = 3(s_{11} + 2s_{12}).$$

In theory the coefficient S could be used in conjunction with the bending and twisting moduli to determine s_{11}, s_{12} and s_{44}, but, in practice, S is but little used because of the experimental difficulty of finding the compressibility.

The linear compressibility is generally much easier to determine (see paragraph 11.4), and if single crystals of reasonable size are available it is better to make measurements of linear rather than of cubic compressibility. The value of the linear compressibility of a non-cubic crystal in any direction may be obtained from the expressions

$$r'_{33} = \alpha_{i3}\alpha_{k3}r_{ik}, \qquad r_{ik} = s_{lmik}t_{lm}$$

when proper values are inserted for α_{i3}, α_{k3}, s_{lmik} and t_{lm}.

11. Experimental methods.

Both static and dynamic methods have been developed for measuring elastic moduli. In the static methods a constant force or stress is applied, and the consequent deformation measured. In the dynamic method the crystal is set into mechanical vibration, either by a blow from a hammer, or by piezo-electric oscillations. The frequency of the oscillations in the crystal is characteristic, and by combining its value with other measured quantities one may deduce the elastic moduli.

The apparatus required for the static methods is relatively standardised though it may be expensive. The displacements to be measured are invariably small, and many types of optical lever and interferometer have been used to measure them. The optical levers can be made self-recording by allowing the emerging beam of light to fall on a moving strip of sensitive film or paper. The tedious labour of counting the fringes passing across the field of the telescope, when an interferometer is used, may be avoided by the use of photoelectric cells.

The elastic moduli may also be determined statically by means of X-rays, using a 'back-reflection' in the same way as described on p. 27, in connection with thermal expansion. The change in direction of the reflected beam, due to the elastic alteration in the size of the cell, may be recorded either by means of an ionisation spectrometer or photographically.

The dynamic method is particularly valuable when the static methods might give rise to plastic deformation, e.g. with perfect crystals of zinc. If a rod or plate of considerable size is available, a musical note of definite pitch is emitted on tapping. The elastic moduli can be calculated from the frequency of this note, and the dimensions of the bar or plate. For substances which cannot be obtained in large crystals, the piezo-electric oscillator is available. A small plate is cemented to the piezo-electrically oscillating plate, the resonant frequency of which is known. The difference in the resonant frequency so introduced is made use of in the calculation of the elastic moduli.

11.1. Static methods of measuring Young's modulus (i.e. $1/s_{33}'$).

Young's modulus Y is defined as follows: If a uniform bar of length L and area of cross-section A is stretched an amount l by a force F, then

$$Y = \frac{F/A}{l/L}.$$

(a) Uniform bending.

If a beam is symmetrically supported on two knife-edges, A and B (fig. 105), and equal forces are applied downwards on knife-edges near the ends of the beam, and at the same distance from the centre of the beam, its curvature between A and B is uniform. If the forces at the ends of the beam are equal to F, then the reaction must also equal F. If the distance between the outer and inner knife-edges is equal to a, then the bending moment B has the value

$$B = Fa.$$

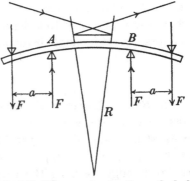

Fig. 105. Diagram showing the method of producing uniform bending in a plate and the method of measuring the radius of curvature by successive reflections from two mirrors.

If Y is Young's modulus for stretching parallel to the length of the beam, R the radius of curvature of the beam between A and B, and I the so-called moment of inertia of the cross-section, the usual relation connecting the bending moment to these quantities is

$$B = \frac{YI}{R}.$$

This relation does not hold exactly for a plate with an arbitrary orientation with respect to the crystallographic axes. A plate cut without regard to the orientation may twist as well as bend under the influence of a simple bending stress; conversely, such a plate may bend as well as twist when subjected to simple twisting. If the formula is to be applied, it is necessary to cut the plate parallel to an axis of symmetry, or perpendicular to a plane of symmetry.

The measurement of R.

(1) *By optical lever.* The only quantity which is at all difficult to measure is R. In the early work this was found by mounting two mirrors between A and B and allowing a light beam to be reflected first from one and then from the other (fig. 105). The change in angle of the direction of the reflected light beam was equal to twice the change in angle between the mirrors, $\delta\theta$. If l was their distance apart, then

$$\delta\theta = \frac{l}{R}.$$

(2) *By interferometer.* R may also be measured by finding the displacement of the centre of the plate. A rod carried in a guide rests at one end on the crystal plate and supports a mirror D at the top (fig. 106). Supported on the same stand which carries the knife-edges bearing the crystal is a transparent plate or lens, C, which may be adjusted so that its bottom surface is parallel to D. Light falling normally† on C passes through to D, and when C and D are properly arranged interference fringes are observed. These are formed by interference of the light

† The departure of the light beams from normal incidence in fig. 106 is exaggerated so that they can be represented distinctly.

reflected by the top surface of D with that reflected from the bottom surface of C. It is advisable to arrange C to be wedge-shaped in order to avoid unwanted interference fringes. If the bottom surface of C has a large radius of curvature Newton's rings are observed. If now the crystal plate be loaded its centre rises and interference fringes move across the field of view. Each fringe comes into the position previously occupied by its neighbour when the displacement of the mirror D is $\lambda/2$. Sodium light or the light from a mercury arc after passing through a filter may conveniently be used. If the number of fringes which have crossed a given point in the field is n, then the displacement of the centre of the plate is

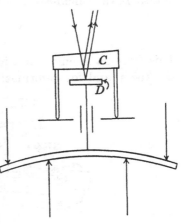

Fig. 106. Diagram showing the method of measuring by the displacement of interference fringes the radius of curvature of the portion of a plate experiencing uniform bending

$n\lambda/2$, and if l is the distance between the inner knife-edges

$$n\frac{\lambda}{2}\left(2R-n\frac{\lambda}{2}\right)=l^2,$$

or, since R is very large compared with $n\lambda/2$,

$$R=\frac{l^2}{n\lambda}.$$

(b) Non-uniform bending.

In the method of uniform bending only that part of the plate between the knife-edges is contributing to the displacement that we are measuring. The method involving non-uniform bending employs the whole plate. In fig. 107 is shown the way in which the fixed and movable plates of the interferometer may be set up and the means of bending the crystal plate.† Two knife-edges A, B at the ends of the plate support it when a force is applied upwards at its mid-point. The displacement of the mid-point s is related to the load P, the length L, the thickness t, and the breadth b, by the formula

$$Y = \frac{1}{4}\frac{PL^3}{st^3b}.$$

This formula is only valid for a bar long compared with its thickness, and for deformation well within the elastic limit.

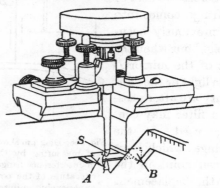

Fig. 107. Showing a crystal plate pressed upwards at its mid-point by a projection on the end of a balance arm against knife-edges, A, B, which are symmetrical with respect to the balance arm. The displacement of the centre of the plate is obtained from the movement of interference fringes formed between the fixed and movable plates. (Reproduced from *Crystallography and Practical Crystal Measurement*, A. E. H. Tutton (Macmillan, London, 1922), by kind permission of the publishers.)

11.2. Static method for measuring the rigidity modulus.

The rigidity modulus is often defined as the ratio of the shearing couple acting parallel to one edge of a unit cube to the change of angle produced between the other two edges.‡

If a rectangular plate of length l, breadth b, and thickness t, is clamped at the two ends and twisted through an angle θ

† For a full account of experimental details and corrections to be applied to the formula for Y reference should be made to *Crystallography and Practical Crystal Measurement*, vol. II, p. 1334, A. E. H. Tutton (Macmillan, London, 1922).

‡ The modulus so defined is common in English usage but is the reciprocal of $4s'_{2323}$ or $2(s'_{2323} + s'_{3131})$ (see p. 246).

by a couple G, the relation between the above quantities and the rigidity modulus n is

$$G = \frac{n\theta}{l} \cdot \frac{bt^3}{3}.$$

This formula only holds if the thickness is small compared with the breadth and in all practical cases requires certain corrections (see footnote, p. 260). The same limitation applies to this formula as to that involving Young's modulus, namely, the crystal plate must be cut parallel to an axis of symmetry or perpendicular to a plane of symmetry.

The measurement of θ.

In the most widely used method the horizontal crystal plate is fixed at one end, and attached to an axle capable of rotating at the other. The torque is applied by rotating this axle. The amount the crystal twists is registered by two mirrors, which are fixed to the ends of the crystal plate, with their planes parallel to its length. The angle through which they turn relative to one another is determined by the 'lamp and scale' method.

11.3. Dynamic methods of measuring Young's modulus and the rigidity modulus.

Bars vibrating freely after impact.

The vibrations which are caused by tapping a crystal rod, when suitably suspended by threads, depend on the manner and direction of tapping, the shape of the bar, and the elasticity moduli. For example, a rod, long with respect to its other dimensions, will vibrate parallel to its length if the rod is suspended at the middle and is lightly tapped in the direction of its length. The middle of the bar will be a node, and the ends anti-nodes, so that the wave-length of the mechanical vibration, λ, will be twice L, the length of the rod. If the frequency of the note is ν, then V, the velocity of the longitudinal waves in the rod, is given by

$$V = \nu\lambda = \nu . 2L.$$

Now by the theory of the transmission of longitudinal vibrations in bars

$$V = \sqrt{\frac{E}{\rho}},$$

where ρ is the density, and E is the appropriate elasticity modulus, in this case the adiabatic bending modulus. Thus

$$E = \rho V^2 = 4\rho\nu^2 L^2.$$

The adiabatic modulus is not quite the same as Young's modulus because the vibrations occur so quickly that any heat generated by the expansion or contraction has no time to escape. The relation of E to Y may be derived from thermodynamical theory. Usually the difference between the two is small, amounting to a few per cent, and this difference varies from one modulus to another for a given crystal.

Determination of the frequency of the vibration.

The simplest method of determining the frequency of vibration of a bar is by direct comparison with a source of standardised frequencies. If the amplitude is too small to be heard it may be amplified by means of a microphone. In another method which has been used a small magnet is fixed to the crystal bar, and a coil arranged round the magnet. When the bar vibrates an alternating current is set up in the coil. This current is amplified and its frequency determined by heterodyning with a calibrated wave-meter. The heterodyne note is made audible by a loud speaker, and the frequency emitted by the wave-meter is changed until the beat frequency becomes zero. The reading on the meter then gives the frequency of free vibration of the crystal plate.

Use of piezo-electric or magnetostrictive oscillators.

A piezo-electric plate or a bar of magnetisable material may be set into mechanical oscillation by the application of electric or magnetic fields. When the frequency of the applied field coincides with that of the natural vibration frequency of the quartz or iron rod, resonance occurs. If now a non-piezo-electric or non-magnetic crystal be cemented to these oscillators the frequency of resonance is changed, the amount of the change depending on the size, shape and elastic properties of the extra rod. The mathematical analysis necessary for dealing with the results of this method is considerable, but the experimental technique is well established and relatively simple. The measurement consists in plotting a resonance curve both with and without the extra crystal rod attached to the oscillator. The elastic modulus is deduced from the difference in the frequencies corresponding to the resonance troughs.

Measurement of the rigidity modulus.

Torsional modes of vibration may be induced by tapping bars or rods when suitably suspended by threads or held in clamps. In one experiment this was achieved as follows. A crystal bar was suspended vertically by a long bifilar suspension and provided with an H-shaped mass at each end. The planes of the H's were perpendicular to the length of the rod. On tapping one of these H-shaped pieces, the rod took up torsional as well as flexural vibrations. The outer portions of the upper H were bar magnets, and by studying the currents induced in coils placed over the four poles it was possible to distinguish between the two types of vibration. From the frequency of this torsional vibration and the dimensions of the rod or plate, the corresponding rigidity modulus was calculated.

Piezo-electric plates or rods may be set into torsional oscillation by a proper disposition of electrodes and an application of suitable alternating potentials. The frequency of resonance for such types of oscillation depends on the elastic properties and size of the oscillator. When a non-piezo-electric crystal rod is cemented on to the quartz, the frequency of resonance is changed. No record exists in the literature of this method having been used, but in principle it is a possible one.

11.4. Determination of compressibility.

Early measurements of the compressibility of solids were made in the following way. A liquid was placed in a vessel so constructed that a small change in volume of the liquid could be measured. The vessel was placed in another strong container filled with liquid and subjected to hydrostatic pressure. The apparent change in volume of the liquid under test corresponded to the difference of the compressibilities of the liquid and the vessel holding it. If the compressibility of the material of the container were calculated from measurements of the elastic moduli that of the liquid could be found. The solid to be investigated was now introduced into the inner vessel and thereby displaced a certain amount of liquid. Pressure was applied as before and the apparent change in volume of the liquid in the inner vessel was the resultant of the contributions due to the solid, the remaining liquid and the vessel. By difference the effect of the solid itself could be obtained.

This method was inaccurate, due to the small compressibility of solids relative to liquids. Also, valuable information on the contraction in various directions under hydrostatic pressure of non-cubic crystals could not be obtained. Both of these difficulties were overcome by methods due to Bridgman,[†]

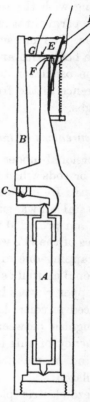

Fig. 108. Apparatus used for determining the linear compressibility of crystal rods. (Reproduced from *The Physics of High Pressure*, P. W. Bridgman (Bell, London, 1931), by kind permission of the publishers.)

[†] P. W. Bridgman, *The Physics of High Pressure*, p. 157 (Bell, London, 1931).

in which the linear compressibility was measured in various directions. The apparatus is indicated in fig. 108. The crystal rod A carried knife-edges at the two ends and rested on a V-groove at the lower end and supported a lever B at the upper end. B pivoted about the knife-edge, C, and magnified the movement of A about seven times. B was held in place by a flat spring D, mounted on the frame of the piezo-meter. Attached to the upper end of B was a resistance wire, E, which could slide over a contact, F, fixed to the frame. The potential between a junction G on E and F when a constant current flowed through the wire was a measure of the position of the top end of B relative to the frame. Thus any change in length of A relative to the frame, when the whole apparatus was put in a liquid and subjected to pressure, was determined by the change in potential between G and F. This method has proved to be satisfactory and most of the measurements at present available have been made by means of it. Care must be taken to choose the orientation of the rod correctly so that on compression it simply contracts along its length. Since deformation under hydrostatic pressure is, as we saw in paragraph 7, homogeneous, the same limitations on the choice of suitable rods apply in the various crystal systems as were discussed in connection with thermal expansion and other properties expressible by second order tensors.

12. The relation between the elastic moduli and the nature of the atoms or ions constituting the crystal.

The compressibilities of the elements in polycrystalline form show a well-marked periodic variation with atomic number. This is true in spite of the large anisotropy in the linear compressibilities in some of the elements. The alkali metals and halogens are the most compressible and the range of values over the Periodic Table is great, e.g. the ratio of the compressibilities of caesium and diamond is 240.

In the alkali halides it is in general true that the larger and more polarisable the ions the greater is the coefficient of cubical compressibility. This is shown in the following table, which gives data for the term a in the expression for the compressibilities between 0 and 12,000 kg./cm.$^{-2}$† (The compressibility κ is expressed in relation to the pressure p by the formula $\kappa = a - bp$.)

TABLE XXXIV. *Coefficients of cubical compressibility of the alkali halides at 30° C. expressed in units of* 10^{-13} cm.^{2}/dyne

	Li	Na	K	Rb	Cs
F	11·7	21·1	33·0	—	42·4‡
Cl	34·1	42·6	56·2	66·4‡	59·4‡
Br	43·0	50·8	67·0	79·3	70·5‡
I	60·1‡	70·7‡	85·3	95·6	85·6‡

The ions of the alkali halides all have completed outer shells of electrons of the rare gas configuration, and regularities of this kind do not appear if we compare the elastic properties of crystals containing atoms or ions of differing electron configuration.

Thus for the face-centred cubic metals we have the following elastic moduli:

TABLE XXXV. *The elastic moduli of some cubic face-centred metals*

	s_{11}	s_{12}	$4s_{44}$
	(× 10^{-13} cm.^{2}/dyne)	(× 10^{-13} cm.^{2}/dyne)	(× 10^{-13} cm.^{2}/dyne)
Al	15·9	− 5·80	35·16
Cu	12·91	− 5·23	16·36
Ag	23·2	− 9·93	22·9
Au	23·3	−10·65	23·80
Pb	93·0	−42·6	69·4

† P. W. Bridgman, *The Physics of High Pressure*, p. 163 (Bell, London, 1931).
‡ Landolt-Börnstein, *Physikalisch-chemische Tabellen, Dritter ergänzungsband* (Springer, Berlin, 1936).

It is clear that since the crystal structure is the same for all these metals, the differences in the ratios $s_{11} : s_{12}$, $s_{11} : 4s_{44}$ must be attributed to the atoms themselves.

13. The relation between the elastic moduli and the crystal structure.

Since the elastic moduli depend so much on the particular atoms or ions present in the crystal great caution must be used in drawing conclusions about the effect of crystal structure on the elastic properties.

The closeness of packing appears to affect the cubic compressibility to a marked extent, as is shown by the following figures:

TABLE XXXVI. *Comparison of the coefficient of cubical compressibility for certain cubic crystals*

Substance	Cubic compressibility	Radius of oxygen ion
	$(\times 10^{-13}$ cm.^{2}/dyne)	
As_2O_3	94·7	1·42
MgO	7·2	1·35
Olivine, $(Mg, Fe)_2SiO_4$	8·1	1·3
Sapphire, Al_2O_3	3·22	1·25–1·36

The structure of arsenious oxide is a very open one, that of magnesium oxide less so, while in olivine and sapphire the close-packing of the oxygen ions is still greater. This is shown by the decreasing radius of the oxygen ion.

The relation between the crystal structure and the elastic properties can be traced in the hexagonal and rhombohedral metals. In Table XXXVII the linear compressibilities† (change in length per unit length per unit hydrostatic pressure) parallel and perpendicular to the hexad axis at 30° C. are given for a number of these elements:

† The linear compressibilities parallel and perpendicular to the principal axis, $K\|$ and $K\perp$, are given by

$$K\| = 2s_{13} + s_{33},$$
$$K\perp = s_{11} + s_{12} + s_{13}.$$

TABLE XXXVII. *The linear compressibilities parallel and perpendicular to the principal axis of a number of metals*

Structure type	Substance	Linear compressibility		Ratio of linear compressibilities $K\|/K\perp$	c/a
		$\|$ hexad$\}$ axis $\|$ triad	$\perp$ hexad$\}$ axis $\perp$ triad		
		($\times 10^{-13}$ cm.^{2}/dyne)	($\times 10^{-13}$ cm.^{2}/dyne)		
A 3	Be	2·25	2·89	0·78	1·58
	Mg	9·9	9·9	1·00	1·63
	Zn	14·88	1·76	8·5	1·86
	Cd	18·7	2·14	8·7	1·89
A 7	As	>17·0	1·89	>9·0	—
	Sb	16·84	5·4	3·13	
	Bi	16·28	6·76	2·41	
A 8	Te	−4·2	28·0	−0·15	—

In structure type A 3 it will be seen that as c/a passes through the value of 1·63, which corresponds to a close-packing of perfect spheres, the ratio of the linear compressibilities changes from less than to greater than unity. Thus the structure of beryllium may be regarded 'as a hexagonal close-packed structure compressed along the hexad axis. The structure of zinc may similarly be regarded as one much drawn out along the same axis.

The structures of arsenic, antimony and bismuth are layer lattices with the planes of the layers perpendicular to the trigonal axis. From analogy with the A 3-type metals we should expect that the ratio of the linear compressibilities, $K\|/K\perp$, would be greater than unity. This is found to be so. Tellurium is the only representative of a chain lattice among the metals given in the table. The value of $K\|$ should be considerably less than $K\perp$, since the atoms are closest together along the chains. In fact, this expectation is more than fulfilled, for the crystal has the remarkable property of expanding along the hexad axis when subjected to uniform compression, as is evident from the negative sign of $K\|/K\perp$.

There is very little available experimental data on the elastic properties of non-cubic compounds. The values of $K_{\parallel}/K_{\perp}$ for calcite, $CaCO_3$, and sodium nitrate, $NaNO_3$, are 2·68 and 3·44 respectively, from which we see that the greater strength of binding in the planes parallel to the CO_3 or NO_3 ions which characterises the other physical properties applies also to the linear compressibility.

Wasastjerna[†] has discussed in detail for the alkali halides the implications of the equation for the potential energy of an ion in a crystal lattice which is usually expressed:

$$\phi_r = \frac{Ae^2}{r} + \frac{B}{r^n}.$$

He showed that the arbitrary character of this assumption must lead to errors and formulated a more complicated expression which gives good agreement between calculated and observed values for all the elastic constants of the alkali halides. This theory only applies, however, to ions having a rare gas electron configuration, which virtually restricts it to structures of the rock-salt type. Further, as is shown in paragraph 15, it is doubtful if any theory based on the assumption that central forces alone determine the elastic equilibrium is applicable.

14.* Special hypothesis about the atomic field of force— Madelung constant.

The considerations of the preceding paragraphs are independent of the magnitudes of the forces which the atoms or ions exert upon one another. We have merely supposed that they are arranged on one or more interpenetrating lattices and that if subjected to small stresses or strains the crystal is stable and returns to its original shape when the forces are removed. The next development of the theory depends upon making certain assumptions about the forces which the atoms

[†] J. A. Wasastjerna, *Soc. Sci. Fenn. Comm. Phys. Math.* 8, 9 (1935); *ibid.* 8, 20 (1935).

exert on one another. It is known that in crystals, such as rock-salt, the atoms are ionised, the sodium being positively charged and the chlorine negatively. The electrostatic forces are assumed to follow the usual Coulomb law, and it is possible to calculate the potential energy of an ion in a crystal lattice without any other *ad hoc* assumptions.

Such an arrangement of positive and negative charges as we have imagined in the previous paragraph is inherently unstable. The potential at the point occupied by a positive charge is negative and therefore there is a tendency for the whole lattice to contract, since this would cause a loss of potential energy. This tendency would lead to a complete collapse of a lattice made up of point charges. We know from many experiments that ions have a finite size which is constant within certain limits. Thus the assumption of simple point charges is valid only for calculating the contribution of the electrostatic forces to the total energy, but the conditions of equilibrium can only be studied if the ions are assumed to have other properties than those of simple point charges.

The potential energy of an ion varies with the distance between the ion and an earthed conductor, being a maximum when the latter is infinitely far off. If the ion, with a charge ϵ, is surrounded by an earthed container of radius r_1, the drop in potential energy corresponds to

$$\int_{\infty}^{r_1} \frac{\epsilon \, dr}{r^2} = -\frac{\epsilon}{r_1}.$$

In a crystal lattice of the rock-salt type positive ions are completely surrounded by negative ions, and *vice versa*; hence it follows that each ion is surrounded by a cubic zero-potential surface. Were sheet conductors to coincide with this surface no effect would be produced on the potential energy of the ions when the conductors were earthed. Hence the drop in potential energy due to the interaction of the ions in a crystal of the rock-salt type may be calculated by treating the ion together with the zero-potential surface as a condenser. This calculation is difficult, but limits are easily set by considering the capacities of the condenser when the ion is surrounded by two spheres, one inscribing the zero-potential surface, and the other circumscribing it. If the distance between positive and negative charges in the crystal is r_0, then the radii of these spheres will be $r_0/2$ and $r_0\sqrt{3}/2$ respectively. The drop in the potential energy will therefore lie between $-2\epsilon^2/r_0$ and $-2\epsilon^2/\sqrt{3} . r_0$, i.e. between $-2\cdot0 . \epsilon^2/r_0$ and $-1\cdot16 . \epsilon^2/r_0$.

This problem was investigated by Madelung, who considered mathematically the mutual effects of an infinite array of positive and negative points. He was the first to show that the potential energy ϕ of any ion in a lattice like that of rock-salt is given by

$$\phi = -1\cdot75\frac{\epsilon^2}{r_0}.$$

Hence $1\cdot75$ is called the Madelung constant for the rock-salt type of structure. There is such a constant characteristic for each type of structure.

15.* Cauchy's relations.

It has been shown (p. 238) that any general elastic constant c_{pqrs} has seven others equal to it which may be obtained by interchanging the suffixes in the following ways:

$pq\ rs$	$pq\ sr$
$qp\ rs$	$qp\ sr$
$rs\ pq$	$rs\ qp$
$sr\ pq$	$sr\ qp$

It will be observed that p and q are always together and, similarly, r and s are always together. It was shown by Cauchy that if the particles composing a crystal be supposed to exert on one another a force depending only on their natures and distance apart, this limitation on the combinations of p, q, r and s is removed. In this case

$$pqrs = psqr, \text{ etc.,}$$

and inserting numerals in place of letters we obtain

$$c_{2323} = c_{2233}, \quad c_{1313} = c_{3311}, \quad c_{1212} = c_{1122},$$
$$c_{1312} = c_{1132}, \quad c_{1223} = c_{2213}, \quad c_{2313} = c_{3312},$$

or, using the abbreviated symbols,

$$c_{44} = c_{23}, \quad c_{55} = c_{31}, \quad c_{66} = c_{12},$$
$$c_{56} = c_{14}, \quad c_{64} = c_{25}, \quad c_{45} = c_{36}.$$

Born† and his collaborators have developed a lattice theory of the elastic properties of crystals from which it follows that whenever all the atoms of a crystal are at centres of symmetry Cauchy's relations should hold. A special case of this arises when the atoms are at points on one of the fourteen Bravais lattices.

A simple test of the truth of Cauchy's relations can be obtained in cubic crystals, since c_{12} should be equal to c_{44}. The following table shows that this is reasonably fulfilled for the alkali halides at room temperature. In other cubic crystals the relations do not hold:

TABLE XXXVIII. *Data providing a test of Cauchy's relations for some cubic ionic crystals*

Substance	c_{12}	c_{44}
	($\times 10^{11}$ dyne/cm.²)	($\times 10^{11}$ dyne/cm.²)
NaCl	1·37	1·28
NaBr	1·31	1·33
KCl	0·81	0·79
KBr	0·58	0·62
KI	0·43	0·42
CaF₂	4·48	3·38

The atoms in cubic metals are at centres of symmetry and should therefore obey Cauchy's relations, but the following table shows they do not:

† Max Born, *Atomtheorie des Festen Zustandes* (Teubner, Leipzig, 1923). A short sketch of this theory is to be found in English in P. P. Ewald, Th. Pöschl and L. Prandtl, *The Physics of Solids and Fluids* (Blackie and Son, London, 1932).

TABLE XXXIX. *Data providing a test of Cauchy's relations for some cubic metals*

Metal	c_{12}	c_{44}
	($\times 10^{11}$ dyne/cm.^{2})	($\times 10^{11}$ dyne/cm.^{2})
Copper	12·3	7·53
Silver	8·97	4·36
Gold	16·6	4·0
Aluminium	6·2	2·8
Tungsten	20·6	15·3

In hexagonal crystals the possibility of writing the four suffixes in any order leads to the relations

$$c_{11} = 3c_{12}, \quad \text{since} \quad c_{66} = \tfrac{1}{2}(c_{11} - c_{12}) \qquad \text{(see p. 244)}$$

and $\quad c_{44} = c_{13}, \quad$ since $\quad c_{23} = c_{13}.$

The following table gives the values of these elastic constants for three hexagonal metals, and shows that Cauchy's relations do not apply:

TABLE XL. *Data providing a test of Cauchy's relations for some hexagonal metals*

Metal	c_{11}	$3c_{12}$	c_{44}	c_{13}
	($\times 10^{11}$ dyne/cm.^{2})	($\times 10^{11}$ dyne/cm.^{2})	($\times 10^{11}$ dyne/cm.^{2})	($\times 10^{11}$ dyne/cm.^{2})
Magnesium	5·65	6·96	1·68	1·81
Zinc	15·9	9·69	4·00	4·82
Cadmium	10·9	11·94	1·56	3·75

The failure of Cauchy's relations shows that the fundamental assumption that central forces alone are responsible for the elastic equilibrium is false. Modern theory[†] shows that for the cubic metals the effect of elastic distortion is to change the energy of the electron cloud as well as the electrostatic energy of the fixed ions. This is the factor which is not taken into account at all by the earlier theories.

† K. Fuchs, *Proc. Roy. Soc.* A, **153**, 622 (1936).

APPENDIX I

A THEOREM ON THE TRANSFORMATION OF TENSORS

It is required to prove that the transformation of second order tensors takes place according to the relation

$$p'_{ik} = c_{li}c_{mk}p_{lm}.$$

We shall prove this for a particular example, because it is rather easier to follow than the general case.

Suppose the relation between two vector quantities A, B is given by

$$A_i = p_{ki}B_k,$$

then

$$A'_2 = p'_{12}B'_1 + p'_{22}B'_2 + p'_{32}B'_3$$

and

$$B'_1 = c_{11}B_1 + c_{21}B_2 + c_{31}B_3,$$
$$B'_2 = c_{12}B_1 + c_{22}B_2 + c_{32}B_3,$$
$$B'_3 = c_{13}B_1 + c_{23}B_2 + c_{33}B_3,$$

where c_{ik} are the direction-cosines defining the transformed axes in their relation to the original axes. Hence

$$A'_2 = p'_{12}(c_{11}B_1 + c_{21}B_2 + c_{31}B_3)$$
$$+ p'_{22}(c_{12}B_1 + c_{22}B_2 + c_{32}B_3)$$
$$+ p'_{32}(c_{13}B_1 + c_{23}B_2 + c_{33}B_3)$$

or

$$A'_2 = p'_{i2}c_{ki}B_k. \qquad \qquad \dots\dots(1)$$

We also have the relation

$$A'_2 = c_{12}A_1 + c_{22}A_2 + c_{32}A_3$$

and

$$A_1 = p_{11}B_1 + p_{21}B_2 + p_{31}B_3,$$
$$A_2 = p_{12}B_1 + p_{22}B_2 + p_{32}B_3,$$
$$A_3 = p_{13}B_1 + p_{23}B_2 + p_{33}B_3.$$

Hence
$$A_2' = c_{12}(p_{11}B_1 + p_{21}B_2 + p_{31}B_3)$$
$$+ c_{22}(p_{12}B_1 + p_{22}B_2 + p_{32}B_3)$$
$$+ c_{32}(p_{13}B_1 + p_{23}B_2 + p_{33}B_3)$$

or
$$A_2' = c_{i2}p_{ki}B_k. \qquad \ldots\ldots(2)$$

Both of the expressions for A_2' in equations (1) and (2) may be arranged as the sum of three terms containing B_1, B_2, B_3. Since B is a vector which has no particular direction, B_1, B_2 and B_3 are independent and the coefficients of each in expressions (1) and (2) must be equal. Hence

$$p_{12}'c_{11} + p_{22}'c_{12} + p_{32}'c_{13} = c_{12}p_{11} + c_{22}p_{12} + c_{32}p_{13},$$
$$p_{12}'c_{21} + p_{22}'c_{22} + p_{32}'c_{23} = c_{12}p_{21} + c_{22}p_{22} + c_{32}p_{23},$$
$$p_{12}'c_{31} + p_{22}'c_{32} + p_{32}'c_{33} = c_{12}p_{31} + c_{22}p_{32} + c_{32}p_{33}.$$

We may solve these three equations to obtain p_{12}', p_{22}' and p_{32}' in terms of the c_{ik}'s and the p_{ik}'s. If we multiply the first equation by c_{11}, the second by c_{21} and the third by c_{31}, we obtain

$$p_{12}'c_{11}c_{11} + p_{22}'c_{11}c_{12} + p_{32}'c_{11}c_{13} = c_{11}c_{12}p_{11} + c_{11}c_{22}p_{12} + c_{11}c_{32}p_{13},$$
$$p_{12}'c_{21}c_{21} + p_{22}'c_{21}c_{22} + p_{32}'c_{21}c_{23} = c_{21}c_{12}p_{21} + c_{21}c_{22}p_{22} + c_{21}c_{32}p_{23},$$
$$p_{12}'c_{31}c_{31} + p_{22}'c_{31}c_{32} + p_{32}'c_{31}c_{33} = c_{31}c_{12}p_{31} + c_{31}c_{22}p_{32} + c_{31}c_{32}p_{33}.$$

If these three equations be added, then because[†]

$$c_{11}^2 + c_{21}^2 + c_{31}^2 = 1,$$
$$c_{11}c_{12} + c_{21}c_{22} + c_{31}c_{32} = 0,$$
$$c_{11}c_{13} + c_{21}c_{23} + c_{31}c_{33} = 0,$$
$$p_{12}' = c_{11}c_{12}p_{11} + c_{11}c_{22}p_{12} + c_{11}c_{32}p_{13}$$
$$+ c_{21}c_{12}p_{21} + c_{21}c_{22}p_{22} + c_{21}c_{32}p_{23}$$
$$+ c_{31}c_{12}p_{31} + c_{31}c_{22}p_{32} + c_{31}c_{32}p_{33}.$$

This result may be derived quite generally by employing the tensor notation as follows:

$$A_i' = p_{ki}'B_k'$$
$$= c_{mi}A_m = c_{mi}p_{nm}B_n = c_{mi}p_{nm}c_{no}B_o';$$

† R. J. T. Bell, *Coordinate Geometry of Three Dimensions* (Macmillan, London, 1912).

thus if we put the subscript $o = k$ (since both go through all values 1, 2, 3),

$$p'_{ki}B'_k = c_{mi}c_{nk}p_{nm}B'_k,$$

and for a particular value of k,

$$\underline{p'_{ki} = c_{nk}c_{mi}p_{nm}.}$$

Conversely, if we exchange dashed for undashed letters,

$$p_{ki} = c_{kn}c_{im}p'_{nm}.$$

Transformation of third and higher order tensors.

If two quantities C_{mo} and D_n are related by a third order tensor q_{nmo} according to the equation

$$C_{mo} = q_{nmo}D_n,$$

then the value of q' when the axes are changed is given by

$$q'_{pqr} = c_{np}c_{mq}c_{or}q_{nmo}.$$

This may be proved in a similar manner to that employed for second order tensors. Thus

$$\begin{aligned}
C'_{qr} &= q'_{pqr}D'_p \\
&= c_{mq}c_{or}C_{mo}, \text{ since } C \text{ is a second order tensor,} \\
&= c_{mq}c_{or}q_{nmo}D_n \\
&= c_{mq}c_{or}q_{nmo}c_{ns}D'_s.
\end{aligned}$$

If we put the subscript $s = p$, which we may do since both have all values 1, 2 and 3, we obtain on equating corresponding coefficients of D'_p,

$$q'_{pqr} = c_{np}c_{mq}c_{or}q_{nmo}.$$

The corresponding proofs for higher order tensors are quite similar to the above.

APPENDIX II

THE EQUALITY OF a_{ik} AND a_{ki} WHEN $k \neq i$

The following application of the law of the conservation of energy shows that for the diamagnetic and dielectric tensors $a_{ik} = a_{ki}$ when $i \neq k$.

If the magnetic moment is M, the field strength H, and the angle between them is θ, then the couple G acting on a diamagnetic body is

$$G = MH \sin \theta.$$

If the direction-cosines of M and H are e_1, e_2, e_3 and f_1, f_2, f_3, respectively, then

$$\sin^2 \theta = (e_2 f_3 - e_3 f_2)^2 + (e_3 f_1 - e_1 f_3)^2 + (e_1 f_2 - e_2 f_1)^2.$$

Hence

$$G^2 = (M_2 H_3 - M_3 H_2)^2 + (M_3 H_1 - M_1 H_3)^2 + (M_1 H_2 - M_2 H_1)^2,$$

where M_i, H_i are the components of M and H respectively. The components of G are therefore given by $G_1 = M_2 H_3 - M_3 H_2$, and G_2, G_3, for which similar expressions hold. Now

$$M_2 = a_{12} H_1 + a_{22} H_2 + a_{32} H_3,$$
$$M_3 = a_{13} H_1 + a_{23} H_2 + a_{33} H_3;$$

hence

$$G_1 = a_{32} H_3^2 - a_{23} H_2^2 + (a_{22} - a_{33}) H_2 H_3 + H_1 (a_{12} H_3 - a_{13} H_2).$$

If we make H perpendicular to X_1 so that the angle between H and $X_2 = \phi$, then

$$H_1 = 0, \quad H_2 = H \cos \phi, \quad H_3 = H \sin \phi$$

and

$$G_1 = H^2 [a_{32} \sin^2 \phi - a_{23} \cos^2 \phi + (a_{22} - a_{33}) \sin \phi \cos \phi].$$

The work done in rotating the sphere through 2π about an axis perpendicular to H and G_2 must be zero unless the diamagnetic body is an unfailing source of energy. Hence

$$\int_0^{2\pi} G_1 d\phi = 0.$$

Since
$$\int_0^{2\pi} \cos^2 \phi \, d\phi = \int_0^{2\pi} \sin^2 \phi \, d\phi = \pi$$

and
$$\int_0^{2\pi} \sin \phi \cos \phi \, d\phi = 0,$$

$$a_{23} - a_{32} = 0.$$

Similarly, it may be shown by making H perpendicular to X_2 and X_3 successively that for all values of i and k $a_{ik} = a_{ki}$ when $i \neq k$.

The analysis for dielectric tensors is essentially similar, but for heat flow and electricity the analysis is complicated, and will not be treated in detail here. For further reference see W. Voigt, pp. 345–357, or Th. Liebisch, p. 135. Briefly, the theoretical result is that if $a_{ik} \neq a_{ki}$ then heat or electricity would flow out from a point source along spiral curves, and not in straight lines. This could be detected in properly designed thermal or electrical conductivity experiments, but careful investigation has failed to detect either phenomenon.

APPENDIX III

DEFINITIONS OF CERTAIN MATHEMATICAL, CRYSTALLOGRAPHIC AND PHYSICAL QUANTITIES

(1) A tensor of second order (or rank) is a quantity a_{ik} which becomes a'_{ik} when the axes of reference are changed from X_1, X_2, X_3 to X'_1, X'_2, X'_3. The relation between the dashed and undashed axes is given by the direction-cosines c_{ik} according to the scheme

$$
\begin{array}{cccc}
 & X'_1 & X'_2 & X'_3 \\
X_1 & c_{11} & c_{12} & c_{13} \\
X_2 & c_{21} & c_{22} & c_{23} \\
X_3 & c_{31} & c_{32} & c_{33}
\end{array}
$$

and it may then be shown that

$$a'_{pq} = c_{ip} c_{kq} a_{ik},$$

where the repetition of i and k on the right-hand side implies that they must in turn and independently be given each of the values 1, 2, 3.

(2) The word 'transformation' is used to describe the change from one set of axes to another. When such a transformation occurs all quantities depending on the choice of axes are said to be transformed from one set of axes to another.

(3) An axis of symmetry of order n is present in a perfectly developed crystal when on rotation of the crystal about this line through an angle of $2\pi/n$ every face comes into congruence with the original position of another corresponding face. These axes may be two-fold, three-fold, four-fold or six-fold and are denoted by the symbols 2, 3, 4, 6 in the notation which is adopted here.

An inversion axis is present when in order to bring a face into congruence with the previous position of another face it is necessary to rotate about an axis through a certain angle and then to reflect in a plane perpendicular to this axis.

Such axes may be of order three, four or six and are denoted $\bar{3}$, $\bar{4}$, $\bar{6}$ respectively.

(4) A plane of symmetry is present in a crystal when each face bears a mirror image relation to another face. The plane of the mirror which reflects the position of the one face into that of the other is the plane of symmetry.

(5) In a perfectly developed crystal possessing a centre of symmetry each line passing through one corner and the centre also passes through a second corner at an equal distance from the centre of symmetry.

(6) A Bravais lattice is one of the fourteen possible arrangements of points in space which have crystallographic symmetry and fulfil the condition that the environment about every point is identical with that about every other.

(7) *Definitions of ray surface and wave surface, as used in this book, and a list of the terms used by various authors to designate a ray surface and an indicatrix.*

Ray surface. If a surface be constructed so that the tangent plane at each point is at a perpendicular distance from the origin proportional to the velocity of propagation of a wave front parallel to the plane, this surface is called a ray surface.

(This definition is preferred to one in which the concept of ray velocity is used, since only the velocity of propagation of wave fronts is directly measurable.)

Wave surface. If a surface be constructed so that each radius vector is equal to the velocity of propagation of the wave front travelling in that direction in the crystal, this surface is known as a wave surface.

Author	Term used to describe the ray surface	Term used to describe the indicatrix
Dana	Wave surface	Indicatrix
Groth (Jackson)	Ray surface	Indicatrix
Houston	Wave surface	—
Johannsen	Ray surface	Indicatrix
Liebisch	Strahlenflache	Indexellipsoid
Pockels	Strahlenflache	Indexellipsoid
Rogers and Kerr	Wave surface	Indicatrix
Tutton	Wave surface	Indicatrix
Wood	Wave surface	—

APPENDIX IV

METHODS OF GROWING SINGLE CRYSTALS OF METALS

There are a large number of techniques employed in the preparation of single crystals of metals for experimental purposes, but they all depend either on the recrystallisation which occurs after deformation and subsequent annealing, or on controlled cooling of a melt.

(1) A polycrystalline rod is annealed by heating to relieve all strain. It is then stretched a small amount—usually from $\frac{1}{2}$ to 4 per cent, so that the grains are plastically deformed. The crystal is then heated for several days at an increasing temperature, finally being heated for a short time just below the melting point. The cooling is carried out very carefully in order to avoid strains being set up. The whole rod, under this treatment, becomes monocrystalline. The disadvantage of the method lies in the fact that the orientation cannot be predetermined.

(2) A crucible containing the molten metal is carefully cooled, and a temperature gradient is maintained in the melt. Provided the temperature gradient exceeds a certain critical value and the rate of advance of the crystallisation front is sufficiently slow, it is generally possible to cause the whole contents of the crucible to grow as a single crystal.

(3) A method particularly useful for the conversion of polycrystalline wires of zinc and cadmium into single crystals is as follows. A length of wire is touched on a hot plate, so that the end melts into a little pool. On lifting up the wire an 'appendix' is formed. The wire is placed in a glass tube in such an atmosphere that the oxide coating which holds the molten metal in cylindrical form is neither destroyed nor grows excessively thick. A furnace maintained at a tem-

perature above the melting point of the metal is slowly drawn over the wire, the end with the 'appendix' entering the furnace first. The rate of growth for a wire about a millimetre in diameter may be arranged to be 1 mm. per minute.

(4) This method is also much used for metals having a low melting point. A crucible containing the metal is maintained at a temperature a few degrees above the melting point. A sheet of mica perforated in the centre is floated on the surface. The diameter of the hole determines the diameter of the crystal rod. A seed crystal having the required orientation is used to touch the melt exposed through the hole in the mica, and then withdrawn at a regular rate. The withdrawal is effected by some mechanical arrangement either driven by clockwork or an electric motor. The rate of growth varies with the size of the rod but may be as great as 1 cm. per minute. The diameters of the rods vary from 0·5 to 5·0 mm. The length of the single crystal wire produced by this method is limited only by the distance the rod can be withdrawn, and it has further the advantage that almost any orientation may be obtained although crystals of zinc with certain orientations are difficult to prepare by this method.

INDEX TO AUTHORS

INDEX TO SUBJECTS

Homogeneous deformation, circular
  sections of deformation ellipsoid,
  40
  under hydrostatic pressure, 265
Hooke's law, 234
Hydragillite, see Aluminium hydroxide

Ideal crystals, 61
Ideal lattices, 55
Impurities, influence on thermoelectric effect, 86
Indicatrix, 128
  circular section of tangent cylinder,
    132
  circular sections of, 132
  definition, 280
  derivation of ray surface from, 132
  principal axes of, 129
Induced magnetic moment and magnetising field, relation between, 97
Inversion axis, 279
Iodine, thermal expansion coefficient,
  32
Iridium, thermal expansion coefficient,
  32
Iron, thermal expansion coefficient, 32
Iron pyrites, thermoelectric properties of, 87
Isosthenic lattices, 34
  cleavage, 58
  magnetic anisotropy, 106
  optical properties, 177
  thermal expansion in, 34
Isothermal surfaces, 65
  form of, 72

Kelvin's axiom, 91
Kirchhoff's law, 113

Latent glide elements, 55
Laue photographs, use in studying
  plastic deformation, 46
Layer lattices, 34
  cleavage in, 58
  conductivity in, 84
  magnetic anisotropy of, 106
  optical properties, 177
  thermal expansion in, 34, 36
Lead, elastic moduli, 266
  optical properties, 169
  thermal expansion coefficient, 32
Lead iodide, relation between structure and double refraction, 178
Lead monoxide, relation between structure and double refraction, 178

Lecher wires, 118
Light scattering from vibrating crystals, 222
Lineage structure, 62
Linear groups, double refraction, 177
Linear ion, 35
Lithium, thermal expansion coefficient, 32
Lithium bromide, dielectric constant,
  123
  compressibility, 266
  molecular refraction, 173
  thermal expansion coefficient, 33
Lithium chloride, compressibility, 266
  dielectric constant, 123
  molecular refraction, 173
  thermal expansion coefficient, 33
Lithium fluoride, coefficient of thermal
  expansion, 30
  compressibility, 266
  dielectric constant, 123
  molecular refraction, 173
  thermal expansion coefficient, 33
Lithium iodide, compressibility, 266
  dielectric constant, 123
  molecular refraction, 173
  thermal expansion coefficient, 33
Lloyd's experiment, 148
Lorenz-Lorentz formula, 173

Madelung constant, 269
Manganese, thermal expansion coefficient, 32
Manganese sulphide, anomaly in
  thermal expansion coefficient, 32
Manganous oxide, anomaly in thermal
  expansion coefficient, 32
Magnesium, glide-twinning in, 52
  linear compressibilities, 268
  optical properties, 169
  plastic deformation of, 45
  thermal expansion coefficients, 32,
    35
Magnesium hydroxide, thermal expansion coefficients, 36
Magnesium oxide, cubic compressibility, 267
Magnesite, dielectric constant, 124
Magnetic anisotropy, in isosthenic
  lattices, 106
  in three-dimensional frameworks,
    107
  of aromatic crystals, 107, 108
  of inorganic crystals, 105
  of layer lattices, 106

Printed in the United States
By Bookmasters